T0363978

Tracks, Scats and Other Traces

Barbara Triggs

A FIELD GUIDE TO AUSTRALIAN MAMMALS

Revised Edition

OXFORD
UNIVERSITY PRESS
AUSTRALIA & NEW ZEALAND

OXFORD
UNIVERSITY PRESS

Oxford University Press is a department of the University of Oxford.
It furthers the University's objective of excellence in research,
scholarship, and education by publishing worldwide. Oxford is a registered
trademark of Oxford University Press in the UK and in certain other
countries.

Published in Australia by
Oxford University Press
Level 8, 737 Bourke Street, Docklands, Victoria 3008, Australia

First published 1996
Revised edition 2004
Reprinted 2006, 2008, 2009, 2010, 2011, 2012, 2013 (twice), 2014, 2015, 2017 (twice), 2019, 2020
Reprinted 2021 (twice), 2022, 2023

National Library of Australia Cataloguing-in-Publication data

Triggs, Barbara.
Track, scats and other traces: a field guide to Australian mammals.

Rev. ed.
Bibliography.
Includes index.

978 0 19 555099 3.

1. Mammals—Australia—Identification. 2. Animal tracks—Australia.
I. Title.

599.0994

Reproduction and communication for educational purposes
The Australian *Copyright Act 1968* (the Act) allows a maximum of one chapter
or 10% of the pages of this work, whichever is the greater, to be reproduced
and/or communicated by any educational institution for its educational purposes
provided that the educational institution (or the body that administers it) has
given a remuneration notice to Copyright Agency Limited (CAL) under the Act.

For details of the CAL licence for educational institutions contact:

Copyright Agency Limited
Level 15, 233 Castlereagh Street
Sydney NSW 2000
Telephone: (02) 9394 7600
Facsimile: (02) 9394 7601
Email: info@copyright.com.au

Designed and typeset by Derrick I. Stone Design, Melbourne
Printed in China by Sheck Wah Tong Printing Press Ltd.
Cover artwork by Frank Knight

Links to third party websites are provided by Oxford in good faith and for information only.
Oxford disclaims any responsibility for the materials contained in any third party website
referenced in this work.

FOREWORD

Interest in the native birds and mammals of the Australian bush goes back to the beginning of settlement, and there has been a series of books that cater for this interest, from John Gould to Wood Jones, Ellis Troughton and David Fleay. In the last thirty years the knowledge of the mammals, particularly, has greatly increased, but much of this knowledge is not readily accessible to the general public. The increased interest in the native fauna, in bushwalking and appreciation of wild places has led to the need for a book that would enlarge that appreciation of the mammals.

Unlike birds, which are colourful and mostly active in the daytime, mammals are discreet, shy and mostly active at night. Peter Matthiesen wrote a book called *The Snow Leopard* in which he never saw the elusive animal of the title but gave one the feeling of knowing much about it from the subtle signs it left of its presence in the wild mountains of Nepal. I was reminded of that book when reading this new edition of Barbara Triggs' book, *Tracks, Scats and Other Traces*. For many people who visit the wild places of Australia the mammals that live there are an unknown presence, rarely seen. But the signs of the presence are all around, if one can read them. Barbara Triggs has learnt to read the signs they leave; their tracks, their scats, their bones and their homes.

The first edition of *Tracks, Scats and Other Traces* received instant acclaim because it fulfilled an immediate need. Since its publication it has become an invaluable aid to many people when exploring the wild places of this country. In this wholly rewritten, new edition Barbara Triggs has extended the coverage from the south-eastern crescent to the whole of Australia and has greatly expanded the treatment, especially of tracks and skeletal remains. The detailed keys and guides to each section should also help readers to make accurate identifications from the signs observed and in doing so add a new dimension to their appreciation of Australia's wild places and the shy mammals that live there.

Hugh Tyndale-Biscoe

CONTENTS

Foreword   v

Acknowledgements   viii

Introduction   x
    The Plan of this Book   x

Tracks   1
    Key to Tracks   8
    Guide to Tracks   10

Scats   85
    Key to Scats   90
    Guide to Scats   156

Shelters, Feeding Signs and Other Traces   188
    Key to Shelters   190
    Key to Feeding Signs and Other Traces   192
    Guide to Shelters, Feeding Signs and
        Other Traces   193

Skulls, Lower Jaws, Humeri and Femurs   243
    Key to Skulls, Lower Jaws, Humeri and Femurs   251
    Guide to Lower Jaws, Humeri and Femurs   254

Bibliography   332

Index   334

ACKNOWLEDGEMENTS

Many people have helped in the preparation of this book and I want to thank them all. Without their help the task would have been impossible. To all who assisted by collecting samples of scats, contributing photographs, providing information and in so many other ways, my sincere and grateful thanks.

In addition to the help I received in collecting material for this book, I also have not forgotten the enormous contribution made by all who provided material, help and advice for my first book, *Mammal Tracks and Signs.* Much of the material gathered for that book has been used again in this new version; my sincere thanks to all concerned.

Almost all the new material on tracks was obtained from two sources. I am grateful to the staff at the Healesville Sanctuary, particularly Merrill Halley and Fisk, and the animals in their care. I also owe much to Peter Johnson and staff of the Queensland National Parks & Wildlife Service at Pallarenda, and the wonderful collection of macropods there.

Scats and pellets were collected and sent to me from all over Australia and for these I would like to thank Rosemary Booth, Linda Broome, Greg Clancy, Jeff Cole, Graham Farrington, Tony Friend, Nick Gambold, Alan Horsup, Peter Jarman, Peter Johnson, Menna Jones, Alex Kutt, Sarah Lambert, Pip Masters, Garry Mayo, Peter Menkhorst, David Milledge, Ralph Miller, Les Moore, Andrew Murray, Meri Oakwood, Penny Olsen, Alexander Pollock, Lesley Rogers, Martin Schulz, Claire Smith, Richard Southgate, Joe Stelmann, Jacqui Stol, Philip Stott and John Winter. I also thank Brigitta Flick and Hugh Spencer from the Cape Tribulation Tropical Research Station, Trudy Richards from the Trowunna Wildlife Park, and Greg Parker and Nicole O'Mara from the Ballarat Wildlife Park, who provided scat material which I was unable to collect in the wild. I am also indebted to the late John Barker for scats from captive mammals.

Information regarding shelters, feeding signs and other traces came from many sources, both published and unpublished. In particular, Meri Oakwood and Menna Jones generously provided information about quolls and devils, and Peggy Rismiller contributed much to my knowledge of echidnas. I am most grateful to them and to all who helped in this regard.

For their help in providing information for the distribution maps, my sincere thanks go to Peter Menkhorst (Department of Conservation and Natural Resources, Victoria), Murray Ellis (National Parks & Wildlife Service, NSW), Catherine Kemper (South Australian Museum), Keith Morris (Department of Conservation & Land Management, WA), Owen Price (Parks & Wildlife Commission of the Northern Territory). I also wish to thank Pat Woolley and John Winter for their help in checking maps, and Ronald Strahan for his permission to use material from the first edition of *The Complete Book of Australian Mammals*.

Photographs came from so many people it would be impossible to list them all, but I wish to thank them most sincerely, not only those who have been acknowledged in the appropriate captions but also those whose photos I was unable to use. I also thank them for their patience in the long wait for the return of their material, which was due to circumstances beyond my control.

I particularly want to acknowledge the Photographic Department of the Australian Museum for providing the photographs of skulls and bones. Carl Bento's patience and good humour throughout a trying week was greatly appreciated. I also wish to thank Tim Flannery and Linda Gibson of the Mammal Department for allowing me to use material from the collection, and Lyndall Skillen for her most useful drawings of some of this material.

I would especially like to thank Trisha Wright for her superb illustrations and Lindsay Addison for his excellent photographs of scats and pellets.

I am especially indebted to the late John Seebeck for his review of the draft of the skulls and bones section, and for his advice throughout this long project. Many of the corrections in this edition are based on the errors and omissions detected by John's keen eye for detail.

INTRODUCTION

The tracks, scats and other traces of animal activity are not difficult to find. Look at the ground as you walk along any dusty bush track or muddy stream bank, step off the track into the forest and study the forest floor and the trunks of trees around you, leave the road anywhere in Australia's dry inland and be amazed at the variety of marks in the sand; pause as you ski along a cross-country trail long enough to examine those patterns in the snow — in all these and many other places you will find another dimension, a new and complex mammal world.

The only way to become skilful in finding these tracks and traces is to go out and look for them. This book is designed to help you interpret them — to put a name to the blurred footprint, the pile of old scats, the scratch on a tree trunk, the whitened skull lying in the grass.

Many of Australia's unique and beautiful mammals are not easy to find in the wild. Most of them are nocturnal and extremely wary of humans. Nevertheless, it is possible to gain much information about them simply by understanding the signs they leave behind them. For early man this knowledge was essential for success in hunting and in defence against enemies. In modern society a knowledge of animal signs can still be extremely useful, and it is a vital factor in any study of our shrinking natural environment. Being able to 'read' tracks and signs adds to our appreciation of this environment and, for many of us, it can be both exciting and enjoyable. There's a whole new world out there!

THE PLAN OF THIS BOOK
This field guide is divided into four sections:
>Tracks
>Scats
>Shelters, Feeding Signs and Other Traces
>Bones

In each section there are three parts: a general description to give you basic information about that particular sign, followed by a key designed to lead you to specific text in the Guide that follows and to an identification of the sign you have found. At the end of each Guide there are descriptions of some of the signs of other animals such as birds, reptiles and invertebrates, which can be confused with those of mammals.

Although in the Keys the mammals are arranged according to criteria such as size and shape, rather than in systematic zoological classification order, in the first three of the Guides they are grouped into basically the same categories as the conventional classification system. Thus, in these Guides the mammals are divided into three major groups: the monotremes, marsupials and eutherians (placental mammals). These large groups are further divided into smaller groups.

MONOTREMES
Monotremes are egg-laying mammals. They suckle their young on milk which is secreted through many small openings on the mother's abdomen. The name monotreme means 'one hole', and refers to the cloaca, the single opening for the products of the reproductive system, the digestive system and the kidneys.

There are two families of monotremes, represented in Australia by the Short-beaked Echidna and the Platypus.

MARSUPIALS
Marsupials are a sub-class of mammals in which young are born in a very undeveloped state. After birth they crawl to the mother's pouch, or marsupium, where they attach themselves to the teat of the milk glands.

The marsupials can be broadly separated into three categories: the herbivorous mammals, the omnivorous mammals and the carnivorous mammals.

These are then further divided into smaller groups, and in the Guides I have generally followed the same groupings as the scientific 'families' of marsupials. For those readers wanting further information on the zoological classification system, I recommend *The Mammals of Australia*, edited by Ronald Strahan. This and other references are listed in the bibliography at the end of this book.

EUTHERIANS OR PLACENTAL MAMMALS
Eutherians are a sub-class of mammals in which the young develop in the uterus of the mother, where they are nourished by a placenta. They are born in a more highly developed state than the marsupials, and after birth are provided with milk secreted from the mother's nipples.

This very large group of mammals is also divided into categories (the 'orders' of the zoological classification system) and I have roughly followed these in the Guides. The order Rodentia, for example, includes all rates and mice, but in this book it is more convenient to further divide them into three groups, treating hopping-mice and water-rats separately from other mice and rats because of their very different tracks. The European Rabbit and Brown Hare both belong to the same order (Lagomorpha), and the Cat, Dog and Red Fox all belong to the order Carnivora. There are two orders of hoofed mammals (Perissodactyla and Artiodactyla), and the marine mammals mentioned in this book all belong to the order Pinnipedia. All bats are in the order Chiroptera.

NAMES
The common names of mammals can be confusing, as people living in different areas often have different names for the same animal. Also, when the ful name is used, such as Common Wombat, both names are capitalised, but when wombats are referred to in general terms there is no capital.

The scientific name of an animal gives first the genus name, with an initial capital letter, and then the species name, without a capital. For example, the two species of potoroo, the Log-nosed Potoroo and the Long-footed Potoroo, are both in the same genus, Potorous, and their scientific names are *Potorous tridactylus* and *Potorous longipes* respectively.

The common and scientific names of mammals used in this book are based on Volume 5 of the *Zoological Catalogue of Australia*, published in 1988, with a few exceptions. These include some, such as Spot-tail Quoll instead of Tiger Quoll, where the new name gives a more accurate description of the animal, and others, such as *Petrogale concinna* – the Nabarlek – because the animal has been reclassified into a   different genus since 1988 (this species was *Peradorcas concinna*). Some new names have been added e.g. the Brown Antechinus is now considered to be four species – Brown Antechinus, Agile Antechinus, Dusty Antechinus and Subtropical Antechinus

DISTRIBUTION AND HABITAT

The distribution maps in this book are placed in the Scats Key, as the common names and scientific names are listed in this section, but they should be used when using other signs as well as scats to help in identification.

The maps in this book are given as only broad guides to the present distribution of the mammal species. Little or no data is available for some areas of Australia, and the range boundaries shown on the maps are only approximate. As habitat plays such an important part in distribution, it must be remembered that an animal will not necessarily be found continuously throughout the range shown on any map of this kind. Some species are only found in moist gullies, for example, and others are confined to areas where there are rocky outcrops or caves, and yet they might be marked as present in a large area of the map.

The brief description of habitat given with each map is an indication of where that species is best able to live and reproduce; it does not mean that the animal is not found in other habitats. Some mammals are able to live in a variety of environments, but many favour particular categories of vegetation or landforms. Knowing a mammal's needs and preferences is of great assistance in determining whether or not it is likely to be present in an area. It can be most useful in identification when used in conjunction with the broad distribution of the species.

TRACKS

Reading tracks in the wild is not always easy. The 'perfect' track drawings in this book may not look like the ones you are trying to identify, because only rarely will you find a perfect track. Often only part of a footprint will be found, but it is one of the fascinations of tracking that, with a little practice, you can read tracks from these incomplete prints and sometimes from just a few claw marks. Knowing the shape of an animal's 'perfect' track will help you to use the clues you find in order to identify imperfect tracks.

Footprints will be different on different surfaces — the same animal walking on mud, firm sand and soft sand will not leave the same print (Figure 1) — and irregularities of the ground may alter the shape of the print. Weather can sometimes affect tracks — wind and rain will blur them, and tracks in snow are distorted by a warm sun — but on some surfaces tracks are often recognisable days after they have been made, and it is possible to estimate how old a track is by the amount of weathering it has undergone. Other variations are caused by the gait and speed of the animal.

Figure 1   Tracks of brushtail possum: in mud(left), in firm sand (centre), in soft sand (right)

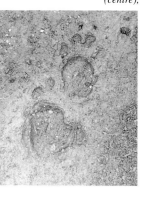

1

The size of the track depends on the age of the animal and sometimes on its sex. In some species the adult females are smaller than the males.

In most cases it is very difficult to distinguish the tracks made by closely related species, unless there is a pronounced difference in size. For example, most rodents have similar feet and therefore leave similar tracks, so even with the clearest tracks it would be difficult to identify which particular species of rat or mouse made the tracks. The use of other signs, such as scats, may help you to limit the choice, and the distribution maps and descriptions of preferred habitat may sometimes enable you to identify an animal accurately to species level.

WHERE TO LOOK
Very few tracks will be found in forest litter, undergrowth or grassy clearings; you must look for places without vegetation where the ground is soft enough to take an imprint. Muddy places and firm, sandy areas often yield good tracks. Creek and dam banks, roadside dust, dried-out puddles, claypans, beaches and snow-covered ground are all good places to look (Figure 2).

The best tracks are found early in the morning, before the wind blurs them and the sun dries them out. They are also easier to see in slanting light. Later in the day, when the sun is higher in the sky, the details tend to flatten out. In bright overhead light, tracks in sand can hardly be seen.

Figure 2   Kangaroo hopping tracks on a sandy beach

STRUCTURE OF THE FEET

A footprint shows the shape of the parts of an animal's foot that touch the ground. Some mammals are *plantigrade*, which means they tread on the whole of the sole of the foot. These mammals usually move at a steady pace; their limbs are relatively short and their feet are not well adapted for running. Wombats, the Koala and the possums are all plantigrade.

Some mammals, like ourselves, are plantigrade when walking, but touch only the toes and the ball of the foot to the ground when running. Animals that run fast and cover large distances generally have longer limbs, and only their toes (digits) touch the ground. These are *digitigrade* animals.

The feet of animals that move mainly by running have evolved so that the heel never touches the ground except at rest. The number of toes of many digitigrade animals has been reduced from five to four — the first toe has either become very small or has disappeared. In the Dog, for example, the first toe is the small 'dew claw' higher up the foot. In kangaroos the front foot has the 'normal' five toes, but in the hind foot the first toe has disappeared, the second and third have been reduced or are   semi-joined, and the fourth is greatly elongated. In hoofed animals the reduction of toes has been even more severe. In deer and other cloven-hoofed animals, the first toe has disappeared and the second and fifth toes have been reduced to dew claws, while in the Horse all the toes have disappeared except the third, the top joint of which is enclosed in the hoof (Figure 3).

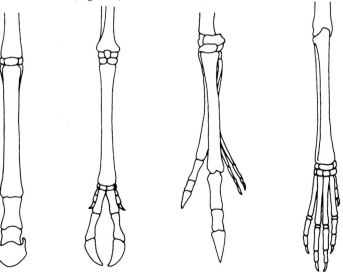

Figure 3   Foot structures of Horse, deer, kangaroo and Dingo (T.Wright)

3

To identify footprints, it is important to know the structure of the feet of the various animal groups. All Australian native mammals, and several introduced species, have 'paws': that is, the feet have claws or nails. In nearly all these animals the soles of the feet have pads, arranged in various ways. It is useful to know the number of pads and how they are arranged, and the number of toes and how many have claws.

Animals without paws have a horny casing (the hoof) enclosing the ends of the toes. It is useful to know whether the species has a single hoof or cloven hoofs.

In many animals with paws, the front and hind feet are different in shape and size, and these differences are most helpful when identifying tracks. The Key to Tracks in this section is based on the shapes and sizes of the tracks, which are all described in the Guide to Tracks.

It can help to learn how to recognise tracks if you make a sketch of any tracks you find, noting their exact size and shape. Photographing them can be useful too, but remember to include a scale or an object of known size in the photograph. Making plaster casts of tracks is another good way of preserving them for further study.

Another way to study tracks is to prepare a smooth area of sand, road dust or mud in a place where you know animals are likely to travel. The good, clear tracks that can be obtained are useful for learning how to recognise tracks, and they are also useful for field biologists studying activity patterns of mammals, their population density and so on.

GAITS AND TRACK PATTERNS

The pattern of a track left by an animal will depend on how fast it was moving and which gait it was using. Few native Australian animals use the standard trot and gallop, and some walk in unusual ways. Some of the useful terms used to describe tracks are:

straddle — the width of the track; that is, the distance between the lines of the left and right feet.

stride — the distance between successive prints of the same foot.

track group or **track pattern** — the distinctive arrangement of footprints made when the animal is using the various gaits.

Walking, trotting and galloping

When an animal is walking, each of the four feet is lifted and set down separately. The prints form two separate, parallel rows. In the walking tracks of Dogs, Red Foxes, Cats and the hoofed animals, the straddle is relatively small, so that these rows are close together, forming a nearly straight line of prints. The hind foot is placed almost exactly in the print of the front foot; that is, the prints 'register', so a two-print track pattern is made (Figure 4). These introduced mammals also use the standard trot and gallop. When the animal trots, the stride is longer and the straddle is narrower than when it walks (Figure 5).

A galloping track shows both hind foot tracks ahead of the two front ones, so a four-print track pattern is made. The faster the gallop, the greater the distance between the individual footprints (Figure 6).

The walking tracks of many Australian mammals have the rows of prints further apart, and the hind foot is placed on the imprint of the palm of the front foot, behind it or beside it. The wider straddle and uneven registration of the prints is characteristic of the walking tracks of the Short-beaked Echidna, Platypus, possums, Koala, wombats and marsupial carnivores. Rodents also have walking tracks of this kind.

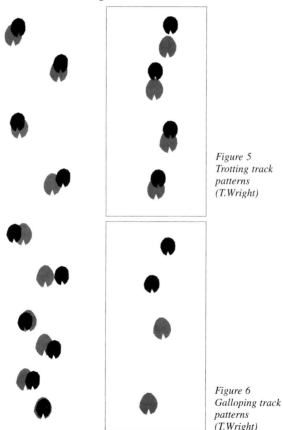

Figure 5
Trotting track patterns
(T.Wright)

Figure 4
Walking track patterns
(T.Wright)

Figure 6
Galloping track patterns
(T.Wright)

Bounding and hopping

When an animal is bounding, it moves by leaping forward with both hind feet and landing with the front feet. Many animals use a half-bound, where the front feet are placed one ahead of the other, and the hind feet land together. The hind feet may land behind, beside or ahead of the front feet. At faster speeds, the front feet may be placed nearly side by side. An animal using a bounding gait leaves a four-print track pattern at each bound. Bandicoots, the Bilby and the introduced European Rabbit and Brown Hare all use a bounding gait, both fast and slow. The faster gait of possums, wombats, the Koala, marsupial carnivores and rodents is also the bound (Figure 7). Some rodents, such as the hopping-mice, use a two-footed (bipedal) hop at fast speeds.

Kangaroos and their relatives have two gaits: bipedal hopping, where the hind feet propel the animal forward and land side by side, and the slow, four-footed gait or 'walk', when the tail is used a fifth 'foot' while the hind feet are brought up beside the front feet. The track pattern is similar to that of a slow bound, with the addition of the mark of the tail (Figure 8).

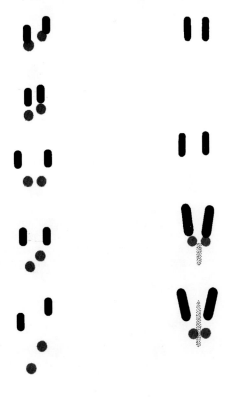

Figure 7
Bounding track patterns
(T.Wright)

Figure 8
Four-footed 'walk' and two-footed hopping track patterns
(T.Wright)

Key to Tracks

STEP 1

Decide whether the prints were made by paws or by hoofs; that is, do the prints show the marks of claws and/or pads (paws), or do they show the sharp outlines of hoofs?

STEP 2

Using the list below and the chart on pages 8 and 9, decide which kind of paw or hoof you have found (A, B, C, D, E, F or G)

PAW PRINTS

Two long, narrow prints with a smaller pair, sometimes with tail drag	A
Two long, narrow prints without a smaller pair	B
Prints of two different shapes and sizes, one with a 'thumb'	C
Prints of two different shapes and sizes, without a 'thumb'	D
Prints of similar shape and size (one may be slightly larger)	E

HOOF PRINTS

Prints of single hoofs	F
Prints of cloven hoofs	G

STEP 3

Measure the length of a single track group or pattern. If a track pattern is incomplete, try to assess how long a complete one would be. The track patterns of the various gaits on pages 5 and 6 will help you to decide the likely positions of the footprints in the pattern you have found. The measurement given below each track is the length of that track, as shown by the arrow beside the first track on page 8.

[Note that the measurements given in the key are for adult animals, and are intended as a general guide only. They do not allow for the smaller tracks of immature animals, nor for the differences in size between males and females in some species. There will also be differences in track length due to the speed of the animal and the nature and condition of the surface.]

KEY TO TRACKS

15–35 cm
pages 15–20

11–19 cm
pages 25–30

A

8–10 cm
pages 31–33

12–15 cm
pages 46–48

B

10–20 cm
pages 15–20

6–13 cm
pages 25–30

4–8 cm
pages 31–33

C

16–20 cm
pages 43–45

10–12 cm
pages 35–36

4–8 cm
pages 37–39

2–6 cm
pages 38–40

12–13 cm
page 11

10–12 cm
page 52

5–11 cm
pages 51–54

1–2 cm
page 55

D

4–5 cm
page 65

1–3 cm
pages 63–64

2–3 cm
page 66

E

7–10 cm
page 70

4–5 cm
page 72

17-25 cm
pages 67–68

6–8 cm
page 71

F

12–15 cm
page 74

12–15 cm
page 74

G

12–15 cm
page 74

10–12 cm
page 75

9–14 cm
page 76

4–8 cm
page 76

4–5 cm
page 76

9

Guide to Tracks

SHORT-BEAKED ECHIDNA

Foot Structure

The Short-beaked Echidna's feet and claws are highly special-ised for digging. The front foot has five strong, broad claws used for digging (Figure 10). The claw on the first toe of the hind foot is short, while the claws on the second and third toes are usually very long (in some subspecies only the second toe is long, and in others the second, third and fourth claws are all long). These long, curved claws are used for grooming between the spines (Figure 11).

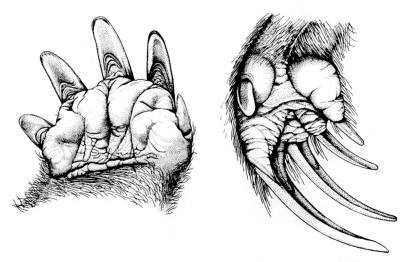

Figure 10   Front foot of echidna
(T.Wright)

Figure 10   Hind foot of echidna
(T.Wright)

Gait

The echidna's usual gait is a slow, rolling walk, with the two legs on one side of the body moving together, followed by the two legs on the other side.

Figure 12
Front foot track
× 1/2

Tracks

Echidna tracks are easy to distinguish from those of other animals. The front feet turn inwards, and the marks of the strong digging claws are usually clear (Figure 12). The hind toes are directed backwards, and the tracks show a print of all the claws as the foot is moved forwards (Figure 13). The long claws of the second and third toes leave a characteristic mark, often leaving drag marks between prints (Figure 15, Plate 1, page 21). When an echidna moves quickly, its stride length increases.

Figure 13
Hind foot track
× 1/2

Figure 14
Walking track pattern

Figure 15
Echidna tracks
in sand

R. MORRISON

11

PLATYPUS

Foot Structure
The Platypus has webbed feet. The front foot has five toes with strong claws, and there is a leathery web between them (Figure 16). This web is folded back when the Platypus walks on land (Figure 17). A small part of the web protrudes under the first claw of the hind foot, but the webbing only reaches the bases of the other four sharp claws (Figure 18).

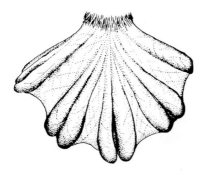

Figure 16   Front foot of Platypus, showing web (T.Wright)

Figure 17   Front foot with web folded back (T.Wright)

Figure 18   Hind foot of Platypus (T.Wright)

Gait
The walking gait of the Platypus is similar to that of the echidna, with both legs on one side moving together. The body and tail drag along the ground. When a Platypus is running the body is slightly lifted, and very little, or none, of it touches the ground.

Tracks
Platypus tracks can sometimes be found on the muddy banks of creeks and rivers, where well-worn pathways lead from the burrows to the water (Figure 21, 22). Often only the imprint of the sharp claws is seen (Figures 19, 20, 22). The five sharp claws of the hind foot curve outwards (Figure 22).

Figure 21
Walking track
pattern on soft
surface

Figure 19
Front foot track × 1/2

Figure 20
Hind foot track × 1/2

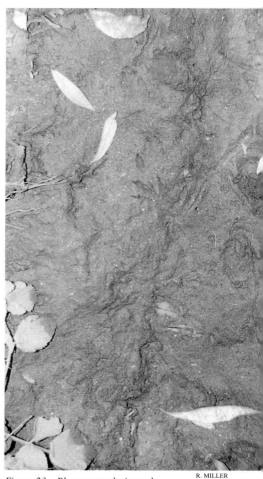

Figure 23   Platypus tracks in mud

R. MILLER

Figure 22
Running track pattern on
firm surface

13

KANGAROOS, WALLABIES AND TREE-KANGAROOS (MACROPODS)

There are about 40 species of kangaroos, wallabies and tree-kangaroos in Australia. They range in size from the large kangaroos, where the males sometimes weigh as much as 65 kilograms, to some small rock-wallabies and hare-wallabies that weigh only 1.5 kilograms.

Foot structure

All kangaroos and wallabies have five clawed toes on the front foot, and these toes point forward (Figure 24). It is the elongated hind foot (Figure 25) this gives this group its family name (macropod = large foot). The long fourth toe, with its large, strong claw, and the shorter fifth toe, which also has a strong claw, are the only ones that touch the ground when the animal moves. The second and third toes are joined up to their top joints and claws, which are separate, These joined toes are used for grooming. The first toe is missing in all kangaroos and wallabies.

Figure 24   Front foot of Swamp Wallaby (T. Wright)

Figure 25
Hind foot of Swamp
Wallaby (T. Wright)

Gait

Kangaroos and wallabies have two different gaits. When moving slowly over a short distance, they have an unusual four-footed gait: the front feet are flat on the ground inside the two hind feet, which are also usually flat on the ground but may sometimes be up on the toes. The front paws are moved forward together; the hind feet then move up together and are placed either side of the front feet. While the hind feet are moving forwards the tail is generally used as a prop (Figure 26), and the mark of the tail appears between or just behind the tracks of the hind feet (see Figures 30, 31).

Figure 26   Walking gait (F. Knight)

In the macropod's faster gait (hopping), the tail and front feet are off the ground. The hind feet are close together, and only the large fourth toe, the smaller fifth toe and the pad behind the toes of the hind feet touch the ground (Figure 27), leaving the characteristic paired track (see Figures 32, 33; Plate 5).

Figure 27   Hopping gait (F. Knight)

TRACKS

Kangaroos, Wallaroos and Large Wallabies

The three kangaroo species, three wallaroo species and seven wallaby species that make up the genus *Macropus*, and the Swamp Wallaby, which is in a different genus, *Wallabia*, all have similar tracks. Of all the native Australian mammals, the tracks of kangaroos and wallabies are probably the ones most commonly found.

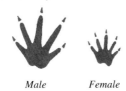

Male   Female

Figure 29
Hind foot track   *Figure 28   Front foot tracks*

15

Figure 30 Walking track patterns of kangaroo (above) and wallaby (below)

Figure 31 Walking track of Eastern Grey Kangaroo in sand

Figure 32 Hopping track patterns of kangaroo (above) and wallaby (below)

Figure 33 Hopping track of Red-necked Wallaby in mud

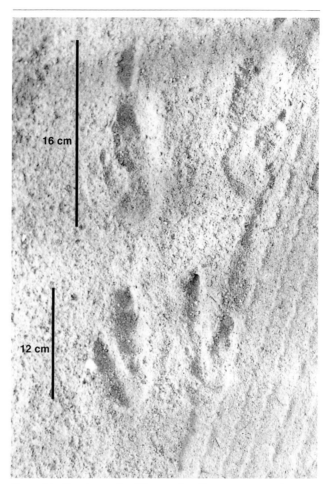

Figure 34   Hopping tracks of Eastern Grey Kangaroo (top) and Swamp Wallaby

The size differences between species of kangaroos and wallabies can sometimes help you to distinguish between the tracks of a large kangaroo and a wallaby, but size is not always a useful guide, as the males of most species are larger than the females. For instance, the footprint of a female kangaroo may be the same size as the print of a male wallaby When clear prints of the front feet show in the track, it is sometimes possible to decide the sex of the animal, as adult males usually have much larger front feet than females. An adult male Eastern Grey Kangaroo's front foot is about 12 cm long; a female's is only about 7 cm long (Figure 28; Plate 4).

The distribution maps for these animals are most helpful. If you find the tracks of a large kangaroo in eastern Victoria, for example, you can be sure they were made by an Eastern Grey Kangaroo, as none of the other large kangaroos is found there. Where the distributions of species of a similar size overlap, as they do in many areas, identification from tracks alone is sometimes possible if a very clear imprint of the hind foot is found.

There is considerable variation in the shapes of the pads on the soles of the feet, and a clear track will show these differences (compare Plates 2, 3, Figures 35, 36, 37).

There is also variation in which parts of the soles actually touch the ground. Some species, such as the Red Kangaroo and the Western Grey Kangaroo, have a moderately dense covering of hair between the pad at the base of the toes and the heel, so that in a print of the whole foot there is a longer gap between the toes and the heel than there is in the Eastern Grey Kangaroo, where this area is naked. In other species the foot is arched, and this prevents part of the sole from touching the ground. If a track shows these pad shapes clearly, it may be possible to decide which species of kangaroo or wallaby made the track (Figure 35).

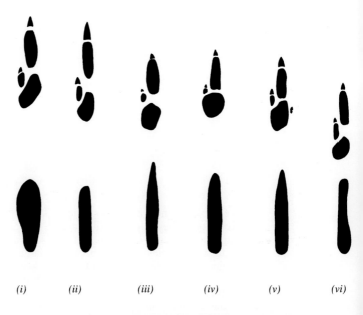

(i) *(ii)* *(iii)* *(iv)* *(v)* *(vi)*

Figure 35   Hind foot of: (i) Antilopine Wallaroo (ii) Western Grey Kangaroo (iii) Common Wallaroo (iv) Red Kangaroo (v) Eastern Grey Kangaroo (vi) Black Wallaroo — all about 1/5 natural size

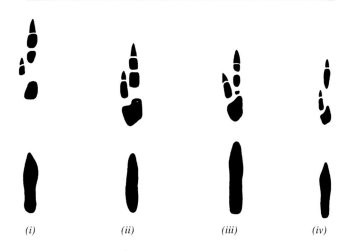

Figure 36   Hind foot tracks of: (i) Red-necked Wallaby (ii) Swamp Wallaby (iii) Agile Wallaby   (iv) Whiptail Wallaby — all about 1/5 natural size

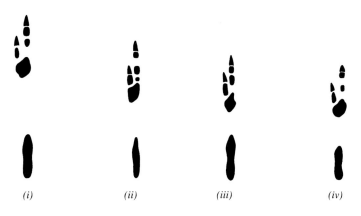

Figure 37   Hind foot tracks of: (i) Western Brush Wallaby (ii) Black-striped Wallaby (iii) Tammar Wallaby (iv) Parma Wallaby — all about 1/5 natural size

19

Figure 38 Wallaby walking tracks in soft sand

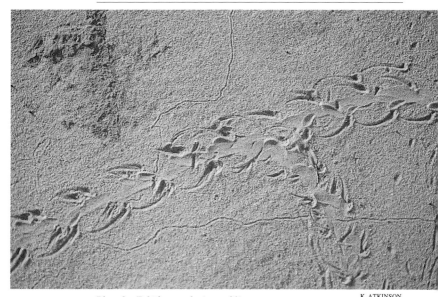

Plate 1   Echidna tracks (page 11)

Plate 2   Eastern Grey Kangaroo hopping track in mud (page 18)

Plate 3   Red Kangaroo hopping track in mud (page 18)

21

K. ATKINSON

Plate 4   Walking track of Eastern Grey Kangaroo (female) (page 17)

D. WATTS

*Plate 5   Red-necked Wallaby tracks
(and Tasmanian Devil tracks) (page 15)*

*Plate 6   Long-nosed Potoroo track in sand
(page 32)*

22

Plate 7 Wombat tracks in firm sand (page 44)

Plate 8 Wombat —marks of claws only (page 45)

D. WATTS

Plate 9 Tracks of Northern Hairy-nosed Wombat (page 44)

23

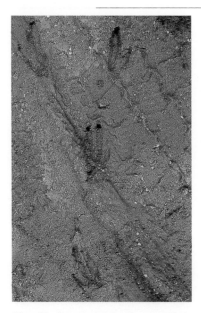

Plate 10   Long-nosed Bandicoot tracks in mud (page 47)

Plate 11   Tracks of Bilby (and Emu) in sand (page 48)

Plate 12   Tracks of Eastern Quoll in snow (page 51)

Plate 13   Tracks of Kowari (and small rodent) in sand (page 52)

Pademelons, Rock-wallabies, Nailtail Wallabies, Tree-kangaroos, Hare Wallabies, Quokka

There is considerable variation in the length and pad shapes of the hind feet of these macropods, which are generally smaller than the kangaroos and large wallabies (Figure 39). If a very clear track is found, these features can sometimes be used to enable a positive identification to be made, with the aid of distribution and habitat information.

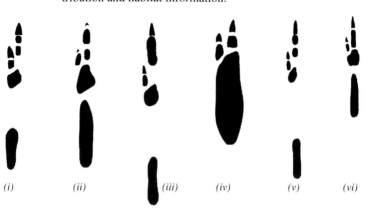

(i)   *(ii)*   *(iii)*   *(iv)*   *(v)*   *(vi)*

Figure 39   Hind foot tracks of: (i) Red-necked Pademelon (ii) Brush-tailed Rock-wallaby (iii) Northern Nailtail Wallaby (iv) Lumholtz's Tree-kangaroo (v) Spectacled Hare-wallaby (vi) Quokka — all about 1/4 natural size

*Figure 40
Hopping track
pattern of
pademelon*

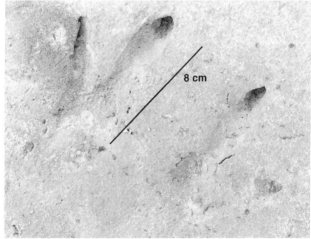

8 cm

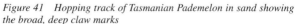

Figure 41   Hopping track of Tasmanian Pademelon in sand showing the broad, deep claw marks

25

Pademelons
The track patterns of these small, compact wallabies are similar to those of the larger wallabies. Often, when they are moving slowly, the weight is transferred directly from the hind legs to the front legs, and the tail simply trails on the ground, leaving a mark behind or beside the footprints (Figure 42). The large, broad claws on the long toes make distinctive, deep marks in the tracks of pademelons (Figures 41, 42, 43).

Figure 42
Walking track
pattern of
pademelon

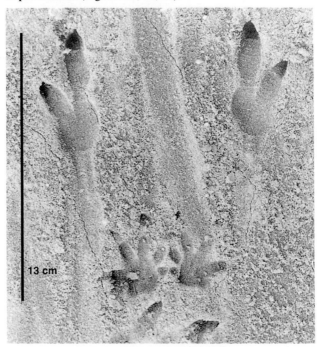

Figure 43   Walking track pattern of Red-legged Pademelon in sand

Rock-wallabies

Broad hind feet with thick, granulated pads and short claws leave the characteristic tracks of the rock-wallabies (Figures 44, 45). The length of the hind foot varies from about 12 cm in the smaller species and females of larger species, to 17 cm in the males of larger species, such as the Yellow-footed Rock-wallaby.

Figure 44
Walking track
pattern

Figure 45   Walking track pattern of Unadorned Rock-wallaby in sand

Nailtail wallabies

The relatively narrow hind feet of the Northern Nailtail Wallaby enable its tracks to be distinguished from the broader hind foot tracks of the wallaroos and large wallabies that share its range (Figures 46, 47).

Figure 46   Walking track pattern

Figure 47   Walking track of Bridled Nailtail Wallaby in sand

Tree-kangaroos

The short, very broad hind feet of these animals leave unmistakable tracks (Figures 48, 49). When on the ground they use the same four-footed walking and hopping gaits as the other macropods, but they also sometimes walk on their hind feet only, moving them independently in short steps, while the long front legs are held close to the body. When climbing trees they walk or run, moving the hind feet alternately, but they also hop along broad, horizontal branches. They often leave footprints of clay or mud on the trunks and branches of the trees they climb.

Figure 48   Hopping track pattern of tree-kangaroo

Figure 49   Tracks of Lumholtz's Tree-kangaroo in sand

Hare-wallabies

These small wallabies have relatively long, narrow feet which leave characteristic tracks (Figure 50, 51). The print of the front foot does not always show in the tracks, but the mark of the tail usually shows.

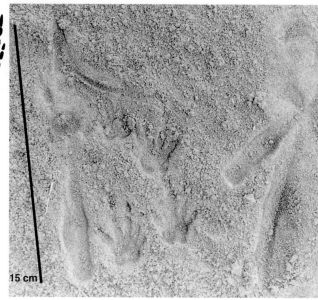

*Figure 50
Walking track
pattern*

*Figure 51
Tracks of
Spectacled
Hare-wallaby in
sand*

15 cm

Quokka

The tracks of these small wallabies show tail marks. The Quokka's hind foot tracks are generally 2–2.5 centimetres longer than those of the recently rediscovered Gilbert's Potoroo, which shares a small part of its range.

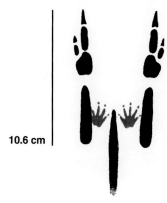

10.6 cm

*Figure 52
Walking track
pattern of
Quokka*

BETTONGS, POTOROOS AND MUSKY RAT-KANGAROO (POTOROIDS)

As with their larger relatives the macropods, the length of the hind foot and the variation in the shape of the hind foot pads are useful characteristics in identifying the tracks of these small animals. Track patterns are similar in shape to the tracks of the larger macropods, as most of these small animals use the gaits — the slow, four-footed walk with the tail trailing on the ground and the fast, bipedal hop. The Musky Rat-kangaroo is the exception – this species does not hop, and it has a first toe on the hind foot.

Bettongs

Although small in body size, bettongs have relatively long hind feet, and their tracks can be confused with those of pademelons (Figure 53, 54).

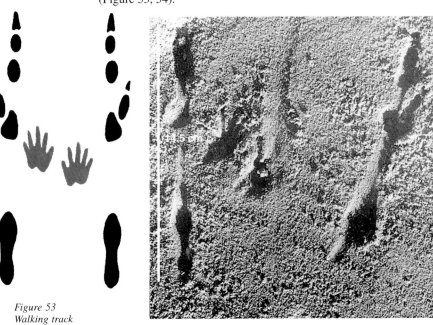

*Figure 53
Walking track
pattern*

Figure 54   Tracks of Rufous Bettong in sand

31

Potoroos

The smaller size of these animals' tracks can sometimes be used to identify them — the Long-nosed Potoroo's hind foot is only about 8.5 cm long (Figures 55, 56; Plate 6). Bandicoot tracks can be mistaken for potoroo tracks, as their hind feet are similar in size and shape, but their front foot tracks are different in shape and are placed differently in the track pattern (Compare Figure 55 with Figures 97 & 99).

Figure 55
Walking track
pattern

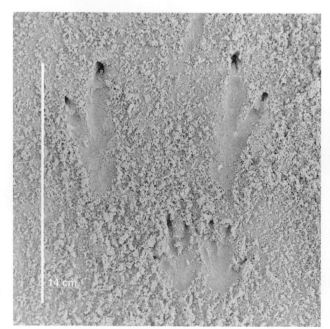

Figure 56
Walking track of
Long-nosed
Potoroo in sand

14 cm

Musky Rat-kangaroo

The Musky Rat-kangaroo does not use the bipedal hop, instead using all four feet in both fast and slow gaits, holding its tail clear off the ground (Figure 58). A clear track will show the print of the first toe (Figure 57) but its tracks are rarely found, as it lives in tropical rainforests.

Figure 57   Walking track pattern

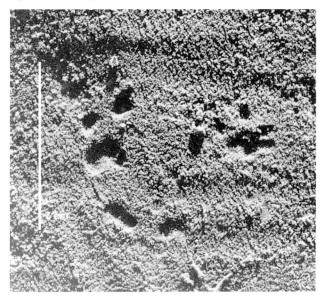

Figure 58   Tracks of Musky Rat-kangaroo in sand

POSSUMS

There are 25 species of possums in Australia. The largest is the Mountain Brushtail Possum, which weighs up to 4.5 kg; the smallest is the Little Pygmy-possum, weighing only about 7 grams.

Foot Structure

The large possums, ringtails, gliders and pygmy-possums all have feet of similar structure. The front foot has five toes (Figure 59) and the toes have strong claws, except in the pygmy-possums and Leadbeater's Possum which have small claws that are often hidden by the pads on the top joints. The front toes are arranged in several different ways. In some species, all the toes point forward; in others, one or two toes are opposed to all the rest. This spread of the front toes is a useful diagnostic feature of possum tracks, and is described in this section under the various species. The Striped Possum's fourth toe is about twice as long as its other toes.

A possum's hind foot also has five toes: a large clawless first toe is opposed to the rest, like the human thumb; the second and third toes are joined up to the first joint (Figure 60).

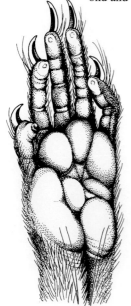

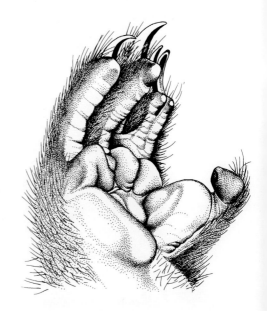

Figure 59   Front foot of Common Brushtail Possum   (T. Wright)

Figure 60   Hind foot of Common Brushtail Possum (T. Wright)

TRACKS AND GAITS

Figure 61
Front foot track of brushtail possum ×1/2

Brushtail possums, Scaly-tailed Possum and Cuscuses

The brushtails often spend some time on the ground during their nightly activities. Their curious, rolling walk, with the hind feet turned out at an angle, leaves an unmistakable track pattern (Figures 63, 64). The hind feet are also turned out in the brushtails' bounding gait (Figures 65, 66).

These possums have the five toes of the front foot more or less evenly spread (Figure 61). The clawless first toe on the hind foot is opposed to the other four toes; the joined second and third toes may leave only one mark on soft ground (Figure 62).

*Figure 62
Hind foot track
of brushtail
possum ×1/2*

*Figure 63
Walking track
pattern*

*Figure 64   Walking tracks of Common
Brushtail Possum in sand*

35

The Mountain Brushtail is larger than the Common Brushtail, and its track is larger and the stride is longer, but it would be very difficult to separate the two species on the evidence of tracks alone.

Cuscuses and the Scaly-tailed Possum have similar gaits to the brushtails and their tracks are about the same size, but their front-foot track differs in that the first two toes are opposed to the other three. Tracks of these animals are rarely found, as the Scaly-tailed Possum lives among rocks and boulders while the cuscuses live in rainforest.

R. MORRISON

Figure 65
Bounding track
pattern

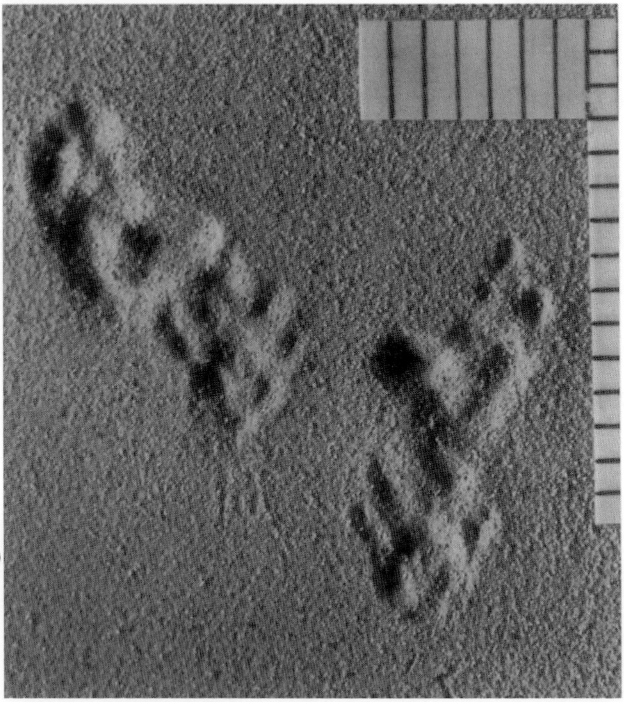

Figure 66   Bounding track of Common Brushtail Possum in sand

Ringtail Possums

Although similar in many ways, it is often possible to distinguish between the tracks of ringtails and brushtails. The ringtails' front toes are spread differently, with the first two opposed to the other three (Figures 67, 69, 70).

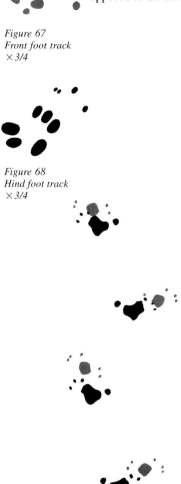

Figure 67
Front foot track
×3/4

Figure 68
Hind foot track
×3/4

Figure 69   Walking track pattern

Figure 70   Walking tracks of Common Ringtail Possum in sand

Tracks of the Common Ringtail Possum are often found, as this species spends some time on the ground, but the rainforest ringtails and the Rock Ringtail Possum rarely leave tracks.

Gliders, Striped Possum and Leadbeater's Possum

Gliders usually travel from one tree to another by gliding and seldom come to the ground, but occasionally their tracks will be found, especially where tree-felling has recently occurred. The usual gaits of these animals are a rolling walk and a bound (Figures 73, 74 and 75).

Figure 71
Front foot track
of Sugar Glider
× 1

Figure 72
Hind foot track
of Sugar Glider
× 1

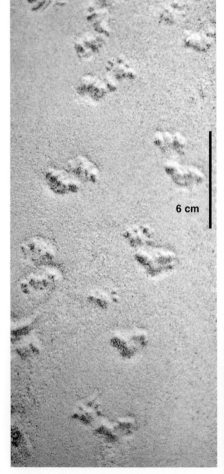

6 cm

Figure 74   Walking tracks of Sugar Glider in sand

Figure 73
Walking track
pattern

38

Although there is much variation in size among the gliders, their hind foot tracks are basically similar. Like other possums, they all have a clawless, opposable first toe on the hind foot, and the second and third toes, which are partly joined, usually make only one mark in the footprint. The prints of the front feet show the difference in the way the toes are spread, and this is a useful diagnostic aid, as it is with the other possums.

The Sugar glider, Squirrel Glider, Mahogany Glider and Yellow-bellied Glider spread only the first toe at an angle (Figure 71). The Greater Glider spreads the two inner toes at an angle to the other three, leaving a similar track to that of the ringtail (Figure 67).

The Striped Possum and Leadbeater's Possum, which are in the same group as the gliders, also spread their toes in the same way as the ringtail (the first two being opposed to the other three). The Striped Possum has an extremely long fourth 'finger', and Leadbeater's Possum has enlarged pads on the top joints of all its toes; although these features would produce distinctive marks, tracks of these possums are rarely found.

Figure 75   Bounding track pattern of Sugar Glider

Pygmy-possums

Pygmy-possums often move about on the ground. Some species are so small and light that their feet rarely leave a mark, but others, such as the Mountain Pygmy-possum and the Eastern Pygmy-possum, are heavy enough to leave an imprint. Their tracks are quite different from those of other small mammals of a similar size. The toes of the front foot of all pygmy-possums are evenly spread (Figure 76), and the hind foot has a comparatively large clawless thumb (Figure 77). Claw marks do not show in the tracks, but the pads on the top joints of all the toes leave clear imprints. Their usual gaits are a rolling walk and a bound (Figure 78).

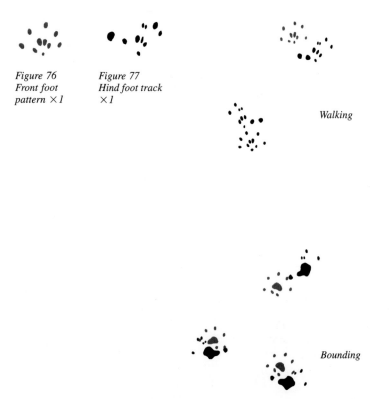

Figure 76
Front foot
pattern ×*1*

Figure 77
Hind foot track
×*1*

Walking

Bounding

Figure 78   Walking and bounding track patterns of Eastern Pygmy-possum

Feathertail Glider

Tracks of these tiny gliders are not often found. They have large pads on the top joints of the toes, and spread the two inner toes of the front foot at an angle to the other three. The hind-foot track is similar to that of the pygmy-possums. Their usual gait when on the ground is a series of jerky jumps (Figure 79, 80, 81).

Figure 79
*Front foot
pattern × 1*

*Figure 80
Hind foot track
× 1*

Figure 81   Jumping track pattern of Feathertail Glider

Honey Possum

The five front toes of this small mammal are evenly spread. The hind feet are unusual, having the fourth toe longer than the others. Its usual gait when on the ground is a fast run, but it rarely leaves an imprint.

KOALA AND WOMBATS

There are three species of wombats and one Koala species, and they are found only in Australia.

Foot Structure

Although the Koala and the wombats all have five toes on the front foot, these are arranged differently. In the Koala the first two toes are opposed to the rest, but in wombats all five front toes point forward (Figure 82). The hind feet have a similar structure to those of the possums: a large, clawless first toe opposed to the other four clawed toes, and the second and third toes joined as far as the first joint (Figure 83).

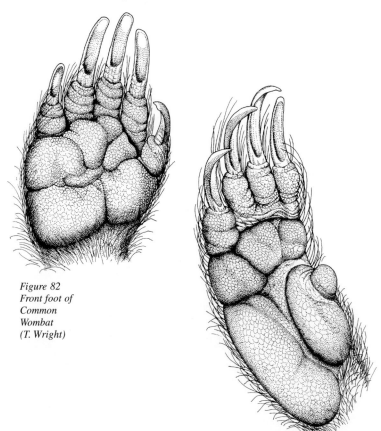

Figure 82
Front foot of
Common
Wombat
(T. Wright)

Figure 83   Hind foot of Common Wombat (T. Wright)

Tracks and Gaits

Koala

Koala tracks may sometimes be found underneath their food trees, as they often come to the ground in order to move from one tree to another. The large front footprint of a Koala is quite distinctive. It shows both the first and second toes jutting out at an angle to the rest of the foot. The sharp, curved claws leave a mark ahead of the rest of the print (Figure 84). The hind-foot print is also large and is similar in shape to that of a wombat, except that the Koala's clawless first toe leaves a more distinct mark (Figures 85, 86, 87).

A Koala's usual gait when on the ground is a sedate walk, but it can move in rapid bounds when alarmed.

Figure 84
Front foot track
× 1/3

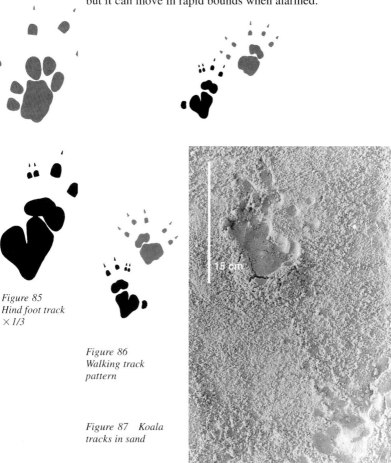

Figure 85
Hind foot track
× 1/3

Figure 86
Walking track pattern

Figure 87   Koala tracks in sand

Wombats

A wombat's track is distinctive and easily recognised. The usual slow, ambling gait and heavy, flat-footed tread result in a curious toed-in track (Figures 89, 90, Plate 7) which has been compared to that of a small, barefoot human (Plate 9).

Figure 88
Front foot track
× 1/2

Figure 89
Walking track pattern

Figure 90   Walking tracks of Common Wombat in soft sand

Figure 91
Hind foot track
× 1/2

Sometimes the hind-foot track is behind or beside that of the front foot (Figure 94), but more often the hind foot registers in the front-foot track.

A slow trot produces a similar track to a walk, but a wombat can move quite rapidly when the need arises, using a bounding gait.

The strong digging claws, particularly those on the front foot, will often leave a clear imprint on firm ground even though the rest of the foot does not leave a mark (Figures 92, 93, Plate 8).

Figure 92
Claw prints

Figure 93   The strong claws leave a characteristic imprint on firm ground

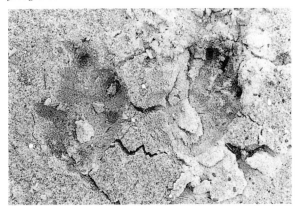

Figure 94
Front (right)
and hind
footprints of
Common
Wombat in sand

L. TREWENACK

45

BANDICOOTS AND BILBIES

There are seven bandicoot species and one Bilby species in
Australia.

Foot Structure

In this group the front feet have five toes but only three have
claws, and these very long claws leave a clear imprint; the first
and fifth toes are small and clawless (Figure 95). The hind foot
is similar to the kangaroo group, except that in the bandicoots
the first toe is present. This toe is very small and clawless and
does not usually show up in the footprint (Figure 96). The first
toe is absent in the Bilby.

*Figure 95
Front foot of
Long-nosed
Bandicoot
(T. Wright)*

*Figure 96   Hind foot of Long-nosed Bandicoot
(T. Wright)*

Tracks and Gaits

Figure 97
Front foot track
× 1/2

Bandicoots

The tracks of the hind feet of a bandicoot can be confused with those of a small potoroo, but a clear imprint of the bandicoot's front foot shows only three toes — the first and fifth do not leave a mark — while the potoroo's shows the usual five toes (Figure 97; compare with Figure 55). Tracks on soft ground may not show the print of the bandicoot's small fifth toe of the hind foot (Figures 97, 98, Plate 10).

Figure 98
Hind foot track
× 1/2

Figure 99
Bounding track
pattern

H. BRUNNER

Figure 100   Bounding track of Southern Brown Bandicoot in firm sand

When a bandicoot moves slowly its gait is a bound, with the front legs moving alternately while the back legs move forward together (Figures 99, 100). A faster bound leaves a similar track, but the front footprints are closer together — they may even be side by side — and the stride is longer (Figures 101, 102, Plate 10). Bandicoots do not hop using only the hind feet as the macropods do.

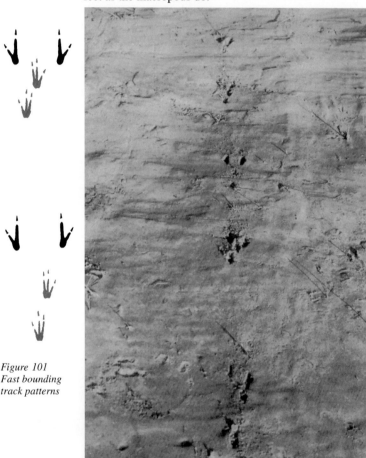

Figure 101
Fast bounding
track patterns

H. BRUNNER

Figure 102   Bounding tracks of Southern Brown Bandicoot in mud

Bilby
Tracks of the Bilby are similar to other bandicoot tracks, but the hind feet turn slightly inwards. The stride of a large male Bilby may be as long as 18 cm (Plate 11).

MARSUPIAL CARNIVORES

There are over 50 Australian species of marsupial carnivores in the Dasyuridae family, commonly known as the dasyurids. They range in size from the large Tasmanian Devil, weighing up to 8 kg, to the tiny planigales and ningauis, which weigh as little as 4 grams. The Thylacine, which weighed about 25–30 kg, is probably extinct on the mainland, but there may still be a small number in forest areas in Tasmania.

The Numbat and the Marsupial Mole are also carnivorous marsupials, although they are not dasyurids.

Foot Structure

All dasyurids have five separate clawed toes on the front foot (Figure 103), but the number on the hind foot varies. All have four clawed toes, and most species also have a small clawless inner toe (Figure 104), but in some species this first toe is absent. Species which do not have this inner toe include the Eastern Quoll, Tasmanian Devil, Kultarr, Kowari and Thylacine.

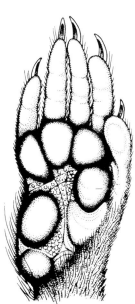

Figure 103
Front foot of
Spot-tailed
Quoll
(T. Wright)

Figure 104   Hind foot of Spot-tailed Quoll
(T. Wright)

49

Tracks

Although the dasyurids vary so much in size, their tracks have many similarities. Their front-foot track may show five clawed toes, generally evenly spread, but the inner toe does not always show in the tracks (Figure 105, 106). The hind-foot track shows four forward-pointing toes, and it is larger and slightly broader than the front foot track and is often turned out at an angle (Figure 105, 107). The clawless inner toe shows in a clear walking track. When the animals are walking, usually only four toes and the pads at the bases of the toes show in the hind-foot track.

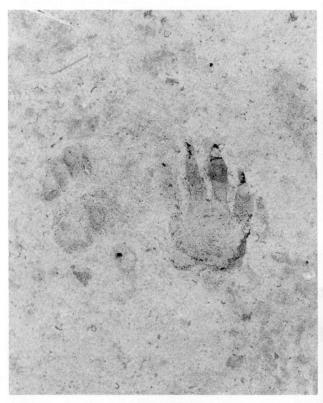

Figure 105   Front foot track (right) and hind foot track of Tasmanian Devil

Quolls, Tasmanian Devil and Some Smaller Relatives

The Spot-tailed Quoll is the largest of the quolls, and its front footprints splay more widely than those of the other quolls, but their track patterns are alike. The small, clawless toe of the hind foot of Spot-tailed, Northern and Western Quolls may be visible in a very clear track.

A quoll's usual gaits are walking and bounding (Figures 108, 109, Plate 12).

Figure 106
Front foot track
of Spot-tailed
Quoll ×1

Figure 107
Hind foot track
of Spot-tailed
Quoll ×1

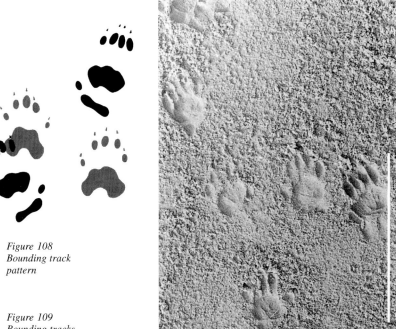

Figure 108
Bounding track
pattern

Figure 109
Bounding tracks
of Spot-tailed
Quoll in sand

51

The Tasmanian Devil has squarish footprints, distinctive because the four forward-pointing toes that are visible on the front and hind foot tracks are evenly spaced and level. The large pads at the bases of the toes leave deep impressions in snow or soft sand (Figure 110). Its usual gait is a slow lope, which leaves a characteristic track (Plate 5).

The Mulgara, Kowari, dibblers and other medium to small dasyurids have similar, but much smaller, tracks (Figure 111, Plate 13).

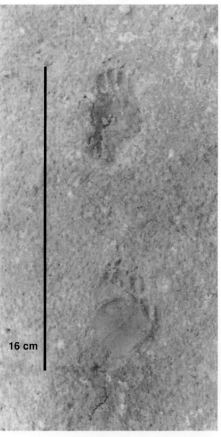

16 cm

Figure 110   Tracks of Tasmanian Devil in firm sand

Figure 111   Bounding tracks of Kowari in sand

R. MORRISON

52

Phascogales and Antechinuses

Although they are largely arboreal, the phascogales also spend time on the ground finding food. They walk when stalking prey (Figure 114), and also bound jerkily (Figure 115).

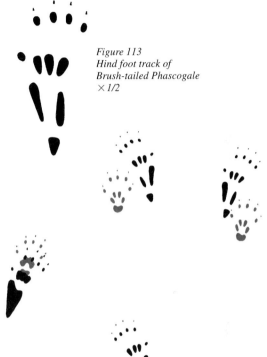

Figure 112
Front foot of
Brush-tailed
Phascogale
× 1

Figure 113
Hind foot track of
Brush-tailed Phascogale
× 1/2

Figure 114
Walking track
pattern of
phascogale

Figure 115   Bounding track pattern of phascogale

The antechinuses are sometimes called 'marsupial mice', but there is no resemblance, other than in size, to rodents. Their tracks are easily distinguished from those of rodents by the arrangement of the toes in the tracks. The five front toe prints are evenly spread, while the four hind toe prints all point forwards (Figure 117; compare this with the Bush Rat tracks on page 63).

The tracks of the smaller Red-tailed Phascogale are similar in size and shape to those of a large antechinus (Figure 119).

*Figure 116
Front foot
tracks of
antechinus*

*Figure 117
Hind foot track
of antechinus*

*Figure 118
Bounding track
pattern of
antechinus*

*Figure 119   Bounding tracks of
Dusky Antechinus in sand*

In a clear track, these animals often show some or all of the outline of the heel, as well as the pads at the bases of the toes. The hind feet are turned out at an angle, leaving characteristic tracks (Figures 118, 119). The tail does not usually leave a mark.

The usual gait of the antechinuses is a series of short, jerky bounds, but they also move slowly at times, using a rolling walk.

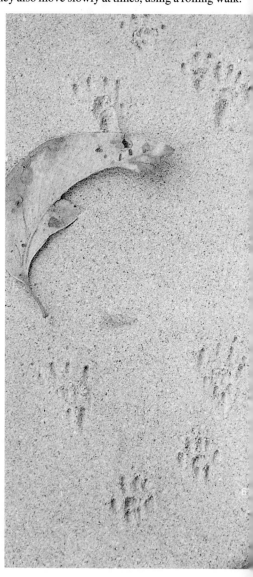

Figure 120
Front foot track
of dunnart

Figure 121
Hind foot track of
dunnart

Dunnarts, Planigales and Other Small Dasyurids

The dunnarts have proportionately longer and narrower feet than the antechinuses, but their tracks and track patterns are very similar (Figures 122, 123).

Figure 122
Bounding track pattern of dunnart

Figure 123   Bounding tracks of
Fat-tailed Dunnart

R. MORRISON

The tiny planigales' tracks also follow the same patterns, but the tail leaves a drag mark in the running track. The Kultarr, in spite of its very long back legs, does not hop; it has a rapid bounding gait.

Figure 124
Front foot track
× 1/2

Figure 125
Hind foot track
× 1/2

Thylacine

If the tracks of the Thylacine were to be found, they would be about the same size as those of a Dog or Red Fox, but with several differences in shape, as the drawings and photographs show. The five toes of the front foot are all short, the inner toe being the smallest; the pad at the base of the toe is large and has a very different shape to that of a Dog or Red Fox (compare Figure 124 with Figures 155 and 156). The hind-foot track shows that there is a larger space separating the toes from the main pad, and this pad is wider but similar in shape to the front foot pad. (Figure 125). Photographs of a captive Thylacine, taken at Hobart Zoo in 1930, show that it sometimes stood on the whole of the hind foot, so that the heel pad would leave an impression (Figure 126), but when walking or bounding only the front part of the foot seemed to touch the ground.

16 cm

L. TREWENACK

Figure 126
Impression of
hind foot in
sand

Plate 14   Dingo tracks on claypan (page 70)

K. ATKINSON

Plate 15   Cat tracks in sand (page 72)

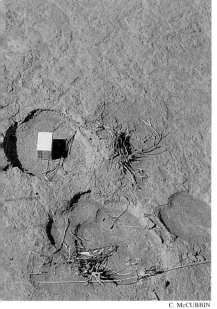

C. McCUBBIN

Plate 16   Camel tracks in sand (page 75)

RISMAC

Plate 17   Sea-lion tracks in sand (page 78)

57

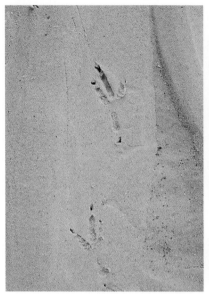

Plate 18   Tracks of raven in sand (page 80)

C. McCUBBIN

Plate 19   Tracks of heron in sand (page 81)

Plate 20   Tracks of Emu in sand (page 82)

K. ATKINSON

58

Plate 21   Goanna tracks in sand (page 83)
Plate 22   Tracks of Saltwater Crocodile (page 83)   ANT / OTTO ROGGE

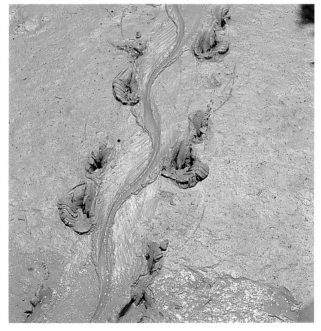

K. ATKINSON

Plate 23   Blue-tongue Lizard tracks (page 83)
Plate 24   Green Turtle tracks (page 83)

K. ATKINSON

Numbat

The Numbat, which belongs to the same order as the dasyurids but a different family, has longer claws on the front foot than does the Western Quoll, and the clawless inner toe is absent, but otherwise its tracks are very similar (Figure 127). The Numbat normally walks and trots, but will bound rapidly if disturbed.

Figure 127   Bounding track pattern of Numbat

Marsupial Mole

The Marsupial Mole is a carnivorous marsupial, but it is probably not related to the dasyurids. In its adaptation to burrowing in the desert sand, two of the front foot claws have become large, spade-like nails. When on the surface, its gait is a kind of shuffling walk. It leaves a track of three parallel furrows made by the legs and short tail. The track ends at a hole where the animal has burrowed into the sand (Plate 99, page 219).

RATS AND MICE (RODENTS)

There are over fifty species of rats and mice in Australia, ranging in size from the Water-rat and Giant White-tailed Rat, which weigh over half a kilogram, to the tiny Delicate Mouse which can weigh as little as 6 grams.

Foot Structure

All rodents have four clawed toes on the front foot and a very small, usually unclawed 'thumb' that does not leave a mark (Figure 128). In all species there are five clawed toes on the hind foot, the inner three of which are considerable longer than the outer two (Figure 129). The Water-rat has partially webbed hind feet. Hopping-mice have narrow, elongated hind feet and the short outer toes do not always leave a mark.

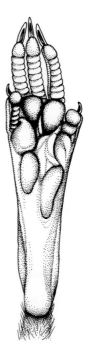

Figure 128   Front foot of Bush Rat (T. Wright)

Figure 129   Hind foot of Bush Rat (T. Wright)

Figure 130
Front foot track
of Bush Rat × 1

Tracks and Gaits

Mice and Rats (other than Water-rat and Hopping-mice)

Except for size variations, the tracks of most mice and rats are extremely similar. The size of a rodent's tracks may give a clue to its identity, and the use of distribution maps is also helpful, but generally it would be impossible to distinguish between them on the evidence of tracks alone. However, it is usually easy to distinguish between the tracks of rodents and those of other small mammals, such as antechinuses.

In a clear track of a rat or mouse, four of the front toes leave a distinctive pattern (one on each side and two pointing forward) (Figure 130), while the larger hind-foot track shows five toes, also spread distinctively (one on each side and three pointing forward) (Figure 131). The heels of the hind feet do not usually leave a mark in the track (Figures 132, 133).

Figure 131
Hind foot track
of Bush Rat × 1

Figure 132   Bounding
track patterns of rat

Figure 133   Bush Rat
tracks in sand

Some of the larger rats, such as the introduced Black Rat, might drag their tail along the ground, and many others sweep the tail on the ground occasionally, leaving an undulating mark in the track (Figure 134).

Figure 134 Black Rat tracks in sand

Walking, running and jumping or bounding gaits are all used by rats and mice (Figures 135, 136).

Walking

Running

Bounding

Figure 135 Track patterns of House Mouse

R. MORRISON

Figure 136 Tracks of House Mouse in sand

Figure 137
Front foot track
×1/2

Figure 138
Hind foot track
×1/2

Water-rat

The front-foot track of the Water-rat shows the marks left by the four long, clawed toes, and is similar to, but larger than, the front-foot tracks of other rats (Figure 137). The partly webbed hind feet leave a distinctive track in soft sand or mud (Figures 138, 139, 140). In firm, wet sand only the five sharp claws leave an imprint (Figures 141, 142).

Figure 139
Walking track pattern

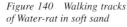
Figure 140   Walking tracks
of Water-rat in soft sand

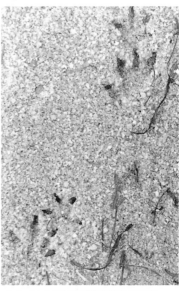

Figure 141
Bounding track
pattern

Figure 142
Bounding track
of Water-rat in
firm sand

65

Hopping-mice

The small, paired tracks made by the hind feet of the hopping-mice can often be found in Australia's inland sandy deserts. These small rodents hop on their hind feet, in the same way as kangaroos, but they also use all four feet in a bounding gait (Figures 143, 144). When moving about slowly, hopping-mice use all four feet in a slow, bounding gait, and the tail leaves a mark in the track (Figures 145, 146).

Figure 143
Bounding track
pattern

Figure 145
Slow bounding
track pattern

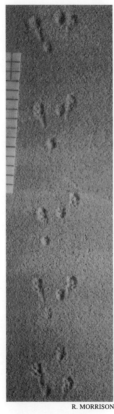

R. MORRISON

Figure 144   Bounding
tracks of Spinifex
Hopping-mouse

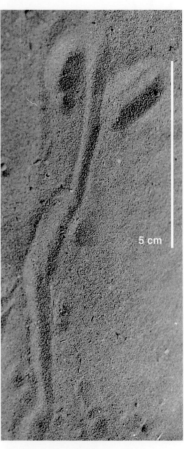

5 cm

Figure 146   Slow bounding tracks of
Spinifex Hopping-mouse in soft sand

RABBIT AND HARE

The European Rabbit and the Brown Hare were introduced into Australia in about the 1860s. Both are widespread pests and the rabbit, in particular, causes severe environmental damage.

Foot Structure
Rabbits and hares have five clawed toes on both the front and the hind foot, but the inner toes are small and leave no mark. A thick layer of hair covers the soles of the feet (Figure 147).

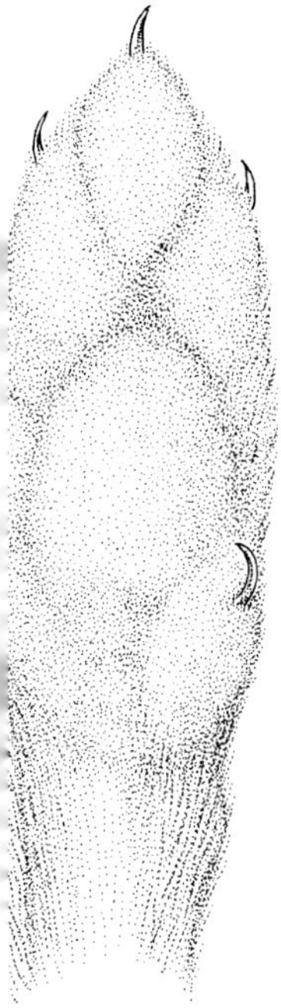

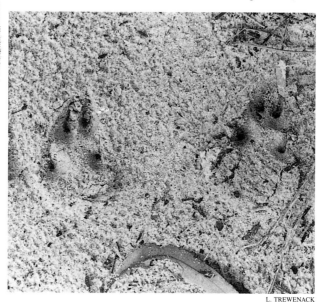

*Figure 147
Front foot of
European
Rabbit
(T. Wright)*

L. TREWENACK

Figure 148   Hind foot tracks of a Brown Hare

67

*Figure 149
Front foot track
of rabbit × 1/2*

*Figure 150
Hind foot track
of rabbit × 1/2*

*Figure 151
Bounding track
pattern*

Tracks and Gaits

The tracks of these animals are alike, but the hare's hind footprint is about a centimetre wider than the rabbit's. The dense covering of hairs on the soles of the feet leave marks similar to those made by pads. Each foot usually leaves four claw and toe marks (Figures 149, 150). On some surfaces, only the claws leave an impression.

The bounding gait of the rabbit and hare leaves an easily recognised track pattern. Each group of footprints consists of two larger hind footprints almost side by side (or with one slightly in front of the other), ahead of the smaller front foot tracks that are almost in a line, one behind the other. (Figures 151, 152). The distance between each group of tracks increases with the speed of the animal, and the distance between each footprint in the group also increases.

When rabbits or hares forage and move slowly, the heels of their hind feet may rest on the ground, leaving a full hind foot track (Figure 153).

R. MORRISON

*Figure 152   Bounding tracks of
European Rabbit in sand*

*Figure 153   Slow bounding
tracks show the full hind foot
foot track*

INTRODUCED CARNIVORES

It is thought that Cats may have been present in Australia for hundreds of years, but many more have become feral since European settlement. The Dingo, which probably came to Australia about 4000 years ago with Aboriginal Australians, and the Dog, which came with European settlers, both belong to the same species. The European Red Fox was introduced in the 1860s and the domestic Cat came with European settlers.

Foxes, feral Dogs and feral Cats are widespread, and are predators on native fauna.

Foot Structure

The Dog, Red Fox and Cat show four toes on each foot in their tracks, with well-defined pads (Figures 154). The front foot has another toe, the first, but it is higher and does not touch the ground. In the Dog and Red Fox, the front foot is larger than the hind foot, but the Cat's are similar in size. All the toes have claws. The Dog's claws are larger than those of the Red Fox. The Cat retracts its claws when walking.

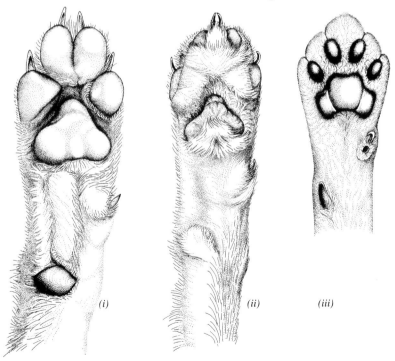

(i)   *(ii)*   *(iii)*

Figure 154   Front feet of: (i) Dog (ii) Red Fox (iii) Cat (T. Wright)

69

Figure 155
Front foot track
× 1/2

Figure 156
Hind foot track
× 1/2

Tracks and Gaits

All these animals walk, trot, lope (a slow gallop) and gallop. They also leap, especially when hunting.

Dog, Dingo

The front-foot tracks of a Dog are larger than the hind foot tracks. The front foot's central pad leaves a wider and clearer print than the central pad of the hind foot, which often leaves only a round or oval depression (Figures 155, 156). There is only a small gap between the print of the central pad and that of the toe pads, in contrast to the fox. The claws on the two middle toes leave clear, well-separated impressions.

When a Dog trots or gallops, the hind-foot track is usually ahead of the front-foot track (Figures 158, 159); in a slow trot, the hind foot often registers in the front track (Figure 157). Dingo tracks are the same as those of a large Dog (Plate 14).

Figure 157
Hind foot track
often partly
registers in front
foot track

Figure 159
Galloping tracks of Dog in sand

Figure 158
Galloping track
pattern

Figure 160
Front foot track
× 1/2

Red Fox

Fox tracks can easily be confused with Dog tracks, but the general shape of the fox's footprint is slightly narrower and more oval than the Dog's (Figures 160, 161). Also, the marks of the two central claws are closer together and usually less distinct — old foxes often have no claws at all.

In the fox, there is a larger gap between the apex of the central pad and the two middle toes than in the Dog. This gap shows clearly in some fox prints, especially in firm sand and snow, where the hair between the toes masks some of the pads. On softer surfaces, such as mud, the gap is not so noticeable, and the tracks are more easily mistaken for Dog tracks (Figure 163).

In fox track patterns the straddle is narrower than in the Dog's; sometimes the prints in a fox's walking track are almost in a straight line (Figures 164, 165).

Figure 161
Hind foot track × 1/2

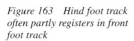

Figure 163   Hind foot track often partly registers in front foot track

Figure 164
Walking track pattern

Figure 162
Trotting track patterns

Figure 165
Walking track of fox in sand

71

Figure 166
Front foot track
× 1/2

Figure 167
Hind foot track
× 1/2

Cat

The Cat's toe pads are arranged almost in a semi-circle in front of the large central pad (Figures 166, 167). The footprints are rounder than those of a Dog or fox. The claws do not usually show in the prints.

The Cat's tracks have a slightly wider straddle than those of a fox, and the footprints are smaller (Figures 168, 169, Plate 15).

Figure 168
Walking track pattern

Figure 169
Walking track of Cat in sand

Figure 170
Bounding track pattern

Figure 171   When a Cat leaps its four feet land close together

HOOFED MAMMALS

There are no native Australian animals with hoofs. All our hoofed mammals have been introduced, mainly for domestic purposes, but many have become adapted to living in the wild. There are six species of deer and eight species of other domesticated or feral hoofed mammals in Australia.

Foot Structure

The hoofed mammals fall into two groups: those with an odd number of toes on each foot and those with an even number. The Horse and the Donkey have an odd number, only one of which — the third — is functional. The small splint bones are the only remnants of the other two toes. The top joint of the large third toe has a strong, almost circular hoof around the pad or 'frog' of the foot, and this is the only part that touches the ground (see Figure 3, page 3).

In the even-toed group, the first toe has been lost. The third and fourth toes are equally well developed, and the tips of these toes form the cloven hoof, which is generally the only part to touch the ground. The second and fifth toes, called dew claws, are high up on the leg (Figure 172). Deer and Pigs sometimes leave an imprint of these dew claws in mud or soft snow (Figures 190, 191, 192). Camels have the third and fourth toes almost completely joined and, instead of forming hoofs, each toe has expanded into a large, semi-circular pad with a nail at the tip.

Figure 172
Front foot of
Sambar Deer
(T. Wright)

Figure 174
Walking track
pattern

Figure 173
Jumping track pattern

Gaits

Hoofed animals, except the Camel, use the standard gaits: walk, trot and gallop; some of them also jump. Generally, walking tracks of those with cloven hoofs show the two toes close together (Figure 174), but in the tracks of faster gaits the toes often splay apart (Figure 173). The Camel has an unusual gait, called pacing, in which the two legs on one side of the body move forward together, followed by the two legs on the other side.

73

Tracks

Horse and Donkey

The hoofs of the Horse leave large, almost circular prints with the V mark of the frog of the foot in the centre (Figures 175, 176). A shod horse leaves a track outlined by the metal rim of the shoe. On firm ground, a hoof print may show only the mark of this metal rim (Figures 177, 178). The hoof prints of the Donkey are smaller than those of the Horse, but otherwise they are the same.

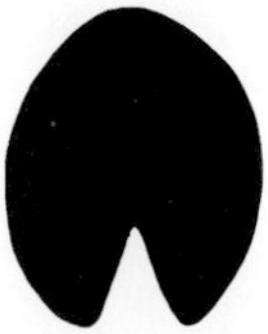

Figure 175
Print of unshod
horse ×1/4

Figure 176
Track of unshod
horse in sand

R. MORRISON

Figure 177
Print of shod
horse

Figure 178   Tracks of shod horse in sand

74

Figure 179
Print of camel
(about 1/5
natural size)

Figure 181
Print of
Domestic Cattle
(about 1/3
natural size)

Camel

Camel tracks are large and easily recognised. They are almost circular, with a small gap between the two semi-circular pads at the front of the foot (Figures 179, 180, Plate 16). The strong nails sometimes show in the tracks.

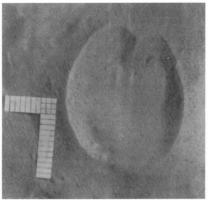

Figure 180   Track of camel in sand (R. Morrison)

Cattle, Buffalo

The large cloven hoofs of these animals leave similar large, rounded prints (Figures 182, 183).

Figure 182   Cattle tracks in sand   *Figure 183   Buffalo tracks in sand*

75

Deer, Sheep, Goat and Pig

Although there is some variation in shape and size between the hoofprints of these animals, it is not easy to distinguish between them. Some are more rounded than others, while some splay apart at the tips more than others.

Figure 184
Print of Fallow Deer ×1/3

Figure 185   Fallow Deer tracks

Figure 186
Print of Goat ×1/3

Figure 187   Goat tracks

Figure 188   Print of sheep
×1/3

Figure 189   Sheep tracks in mud

Figure 190
Print of Hog
Deer ×1/3

The dew claws of Pigs and deer sometimes leave an impression, especially on soft ground; those of deer are closer together than those of the Pig (Figures 190–192).

Figure 191
(Right) Print of
Pig ×1/3

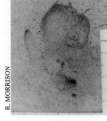

Figure 192
Pig track showing
dew claws

MARINE MAMMALS

There are many species of marine mammals in the seas around Australia. Most spend their entire lives in the water, but seals and their relatives spend part of their lives on land.

Two groups of seals live on or near the Australian coast, or occasionally visit it. The eared seals, which include the Australian Sea-lion and two species of fur-seals, are the only marine mammals commonly found on Australian shores. Of the other group, the true seals, the Elephant Seal is an occasional visitor and four other species make rare visits.

Foot Structure

The front and hind feet of seals have been adapted to form flippers, in which the toes are joined by a web of skin and connective tissue. Sea-lions and fur-seals have long foreflippers, with the bones of the five toes visible under the smooth skin. The first toe is the longest, the fifth the shortest and the claws are small nodules. The hindflippers have long claws on the three middle toes, used for grooming.

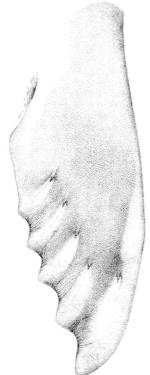

*Figure 193
Foreflipper of
Australian Fur-
seal (T. Wright)*

77

Elephant Seals have shorter foreflippers, with a small claw on each toe; the hindflipper has fibrous tissue between the toes, which increases the flipper's surface area.

Tracks and Gaits

Sea-lions and fur-seals leave similar tracks. The heavy body is supported by the foreflippers, which are extended to the side with a right-angle bend at the wrist, so that the largest (first) toe is to the outside. The 'hands' of the foreflippers leave deep impressions and the marks of the toes sometimes show in the track. The large body leaves a drag mark between the foreflippers, which are moved forward alternately (Figures 194, 195, Plate 17). The hindflippers are also moved alternately, but only the heels rest on the ground. When moving more rapidly, both foreflippers are moved together.

Figure 194
Track pattern of
fur-seal

Figure 195
Tracks of
Australian Sea-
lion (i) going
uphill (ii) going
downhill

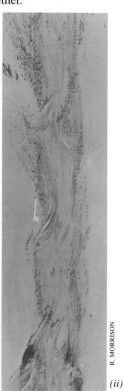

(i)   *(ii)*

R. MORRISON

True seals move on land by humping the body, taking the weight alternately on the chest area and on the pelvic area. The foreflippers are sometimes used to give balance, but the tracks usually show only the wide furrow made by the heavy body.

BATS

There are about sixty species of bats in Australia. They are divided into two main groups: Megachiroptera, or megabats, which includes the fruit-eating flying-foxes and others that feed on nectar and pollen; and Microchiroptera, or microbats, which are insectivorous, except for the carnivorous Ghost Bat.

Foot Structure

The wing of a bat is a greatly modified form of the usual mammal arm and hand. The bone of the forearm is greatly elongated, the 'thumb', or first digit, is short and has a long claw, the second digit is longer (and in the megabats is also usually clawed), and the other three digits are very long. These bones support the thin skin of the wing. This elastic membrane is stretched across the fingertips and is attached along the sides of the body, down the hind leg to the ankle or toes. The hind legs are also joined by a membrane, which usually also encloses the tail (Figure 196).

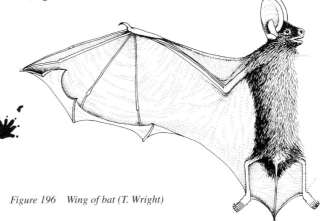

Figure 196   Wing of bat (T. Wright)

Walking track pattern of small bat

Tracks and Gaits

Bats very rarely land on the ground, so it is unlikely that their tracks will be found. Members of one group of microbats, the mastiff-bats, occasionally forage for insects on the ground and can move quite rapidly, and others land accidentally at times.

Bats move over the ground using the same movements that most four-legged animals use when walking, their front and hind limbs moving as diagonally opposite pairs. They use the pad at the base of the 'thumb' to support the forelimb. This leaves a single mark, while the hind-foot tracks show the five toes, with claw marks usually present. The tail tip may also leave a small mark, as this is sometimes used as a point of balance.

79

TRACKS OF OTHER ANIMALS

Many other animals — birds, reptiles and insects, for example — leave tracks that could be confused with those of mammals.

Birds

Figure 197
Paired tracks of
small perching
bird

Perching birds spend much of their time in the trees, but many feed on the ground, where they employ the same hopping gait they use on branches. Their feet have three toes pointing forward and one pointing backward. Wrens, robins, sparrows and many other small birds leave paired tracks due to this hopping gait (Figure 197).

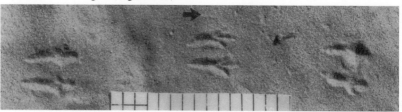

R. MORRISON

Magpies, currawongs, crows and many other perching birds spend a lot of time on the ground. These birds usually walk, leaving alternate tracks (Figures 198, Plate 18) but they also leave paired tracks when they land and take off, and when they hop (Figure 199).

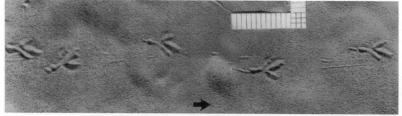

R. MORRISON

Figure 198
Alternate tracks
of larger
perching bird

Figure 199
Landing track
of raven

Cockatoos, cuckoos and birds of the parrot family have two toes pointing forward and two backward. They walk, leaving alternating tracks (Figure 200).

Figure 200
Tracks of Galah

Wading birds have three long, forward-pointing toes, and some, such as herons, also have a long, backward-pointing toe (Figures 201, 202, Plate 19). These birds leave alternating tracks.

Figure 201
Tracks of White-faced Heron

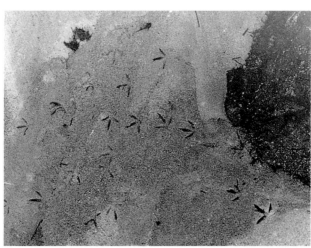

Figure 202
Tracks of small wader, species unknown

81

Ducks, swans and other swimming birds have webbed or partially webbed feet. Their alternating tracks are found mostly on beaches and stream banks, and are easily recognised and unlikely to be confused with mammal tracks (Figures 203, 204).

Figure 203
Black Swan
Tracks

Figure 204
Little Penguin
tracks

P. REILLY

Flightless birds, such as the Emu and Southern Cassowary, walk and run but do not hop. Their three-toed feet leave prints that can be mistaken for kangaroo or wallaby tracks (Figures 205, Plates 11, 20), but the birds leave alternating tracks whereas kangaroo and wallaby tracks arc paired.

Figure 205
Footprint of
Emu

L. TREWENACK

82

Reptiles

Crocodiles and large lizards such as goannas walk and run, leaving alternating tracks. Walking tracks show the drag mark of the body and tail between the alternate footprints on either side (Figure 206, Plates 21, 22). Sometimes the body is lifted off the ground when the reptile is running quickly, but the tail usually leaves a wavy track.

Figure 206
Goanna tracks
in sand

Smaller lizards, such as skinks and geckos, show the imprints of their small, sharp claws either side of a narrow furrow made by the body and tail. Wide-bodied lizards, such as the blue-tongues, leave a wide body scrape, with a narrower furrow made by the tail (Plate 23).

Turtles and tortoises leave distinctive tracks, with the marks of the body between the furrows made by the flippers (Plate 24).

Snakes leave a characteristic narrow furrow, but as there are no footprints these have no similarities to mammal tracks.

Amphibians

Figure 207
Frog track
pattern

The tracks of small toads and frogs are usually hopping tracks; the long fourth toe on the hind foot and the inward-pointing front-foot tracks are characteristic of all amphibian tracks and distinguish them from the tracks of rats and other small mammals (Figure 207).

83

Toad tracks are very common in parts of eastern Queensland and northern New South Wales, where the Cane Toad is a pest. This toad does not always hop; it often walks with the feet dragging slightly (Figure 208).

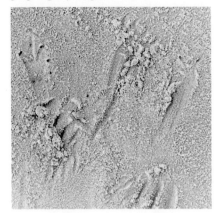

Figure 208
Tracks of Cane
Toad in sand

Invertebrates

Crustaceans such as crabs and yabbies leave tracks which might be confused with mammal tracks. Yabbie tracks, for example, show foot marks on either side of a narrow body scrape, and might be mistaken for small rodent tracks and tail marks (Figure 209).

Figure 209
Yabbie tracks

Most insects are not heavy enough to make an imprint on any firm surface, but some of the larger beetles, crickets and grasshoppers sometimes leave tracks on fine dust or sand. The three pairs of legs leave regular track patterns, often with a drag mark made by the body (Figure 210).

Figure 210
Beetle tracks

84

SCATS

The solid waste material produced by mammals is known by a variety of names — droppings, dung and faeces are some of them, and there are several less polite terms. In this book the word 'scat' is used. It comes from the Greek word meaning excrement or faeces, and is now commonly used to describe this waste material.

The idea of studying scats may be distasteful at first, but they provide so much interesting information that it is worth cultivating an attitude of scientific detachment towards them.

Scats are among the more characteristic signs left by an animal, and they are the ones most likely to be found. They not only indicate which species of animal passed by, and when, but also what it had been eating. They give a guide to the animal's size and weight, its approximate age and even, occasionally, whether it was male or female.

Figure 211 Common Wombats often deposit their scats in prominent places

Scats are sometimes used as territorial markers. Many animals, particularly males, will deposit their scats in prominent places as a sign of territory ownership. Wombats signpost their home range in this way (Figure 211); foxes often leave their scats on top of tussocks or rocks.

Scats are useful indicators of the owners of shelters of all kinds. Animals occupying tree-holes are sometimes identifiable by the scats found at or near the bases of the trees; burrows, caves and hollow logs may also have telltale scats in or near them.

Runways and paths through the undergrowth will often be used by more than one species, and the scats deposited along these trails are useful in identification (Figure 212).

Figure 212
A runway used
by Swamp Rats

For several reasons, the scats of some animals are not readily found. Some scats are so small that they disappear in the leaf litter; some are so friable that they disintegrate almost immediately after deposition. Dung beetles often penetrate and demolish herbivore scats of all kinds. In a few species, such as bettongs, the scats are sometimes collected as soon as they are produced, by dung beetles living on the animal's fur.

Even those scats that are large, long-lasting and easy to find, such as those of macropods, present some problems in identification. For instance, there is considerable variety in the size and shape of pellets produced at one deposit by one animal; there is also variety in these shapes due to seasonal changes in the food — wet or dry grass can alter the consistency of the scats. Another difficulty is the similarity between scats of related species, such as the kangaroos and wallabies, especially where these macropods are eating the same food. Nevertheless, it is often possible to identify the scats of many animals, often to species level, with the help of distribution maps and habitat information.

SCATS OF HERBIVORES AND CARNIVORES

It is usually easy to recognise the difference between the scats of plant-eating animals (herbivores) and meat-eating animals (carnivores). When scats are fresh, the most noticeable difference is their odour. Herbivore scats generally smell like plant humus, a strong but not unpleasant odour. Carnivore scats all have strong, unpleasant smells — some worse than others — and these odours sometimes vary with the nature of the prey eaten. For example, when wild Dogs eat carrion their scats have a particularly offensive smell. 'Learning' smells can be a very useful exercise for those wishing to identify scats of all mammals, but it can only be done by personal experience.

Scats of herbivores are generally dark brown, black or dark green when fresh, but when dry and weathered they become paler. They are produced in large quantities, as much plant material has to be eaten in order to provide sufficient nourishment, and are deposited either as a large mass or in the form of pellets. They are one of the most easily found signs of a herbivore's presence. When a herbivore scat is examined closely the plant material can be seen, and whether this is composed of fine particles or coarser fragments often gives a clue to the identity of the animal.

Scats of carnivores are usually cylindrical, often pointed at one end. There are often fragments of bone and insect material visible, with a twist of hair at one end of the scat. Some scats are twisted or coiled; others are sausage-shaped, often with an outer layer of a white, chalky substance derived from bones that have been digested. Some carnivores are opportunistic feeders — quolls, Dogs and foxes often eat plant material such as fruit and berries.

SCATS OF OMNIVORES AND INSECTIVORES

A number of animals regularly have both plant and animal material in their diet, and are thus called 'omnivores'. Some members of the possum family are omnivores, as are many rodents.

Bandicoots, potoroos, bettongs and the rat-kangaroo all eat insects as well as plants.

The scats of insect-eaters (insectivores) have a characteristic smell, similar to but not as strong as the smell of some carnivore scats. They are usually easily broken, and often small fragments of insects can be seen on the surface. Echidna scats contain large amounts of soil as well as insect remains. Numbats also ingest soil and wood fibres when collecting termites and ants. Most small bats are insectivores.

SCAT ANALYSIS

Detailed examination of scats — scat analysis — yields much useful information about an animal's diet. In particular, the kinds of plants and animals that have been eaten can be identified.

The plant material in a herbivore's scat is analysed by first washing the plant material in the scat, then soaking it in chemicals to soften it. It is examined under a microscope, where its appearance — the pattern of the cells and stomata, or pores — is compared with samples of known plant cuticles (Figure 213). A knowledge of the plant species growing in the area where the scat was collected is very useful for identifying the plant material in the scat.

Figure 213 Microphotograph of some plant cuticles found in a brushtail possum scat

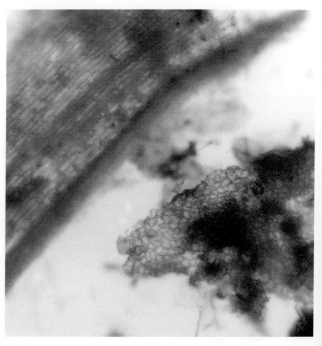

Also, sometimes the identity of the animal itself can be determined, as most animals lick their fur, swallowing some of it and passing this in their scats. By breaking up the scats and extracting the hair or hairs present, it is possible to identify the animal. A comprehensive description of hair analysis is given in *The Identification of Mammalian Hair* by H. Brunner and B.J. Coman (Inkata Press, 1974).

Hair is also present in many carnivore scats, along with fragments of bones and teeth, feathers, reptile remains and insect material. All of these items can be identified by experts. Some of the bones and teeth can be used to identify prey, at least to family level, by using the Guide to Bones in this book.

The regurgitated pellets of birds, some bird scats and reptile scats, illustrated in the Scats section, are included because they can be readily confused with mammal scats (see Plate 56, page 155, and figures 229, 230, 231).

WARNING: Many animals carry worms and other parasites harmful to humans, and the careless handling of scats can result in infection. Hydatids, a disease in which cysts form in the liver and other organs of humans and some other animals, such as sheep and wallabies, is caused by the larvae of certain tapeworms. The eggs of these tapeworms may be present in the scats of Dingoes and Dogs, both wild and domestic, and in foxes, so their scats should be handled with great care and attention to hygiene.

Figure 214
Contents of a
fox scat

Key to Scats

In the following illustrations, the scats are grouped according to their shape and size.

What to do when you find a scat

Step 1   Decide the shape of the scat: is it cylindrical; or is it a pellet, or one of a group of pellets, of oval, round or squarish shape? (Some species produce scats that are of mixed shapes, such as oval and cylindrical, so that they are not easy to classify. As a result, there is some overlap in the Key.)

Step 2   Measure the size of the scat: use the approximate width of the pellet/s or cylinder. If you have several scats, take the average of their widths.

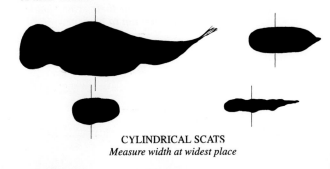

CYLINDRICAL SCATS
Measure width at widest place

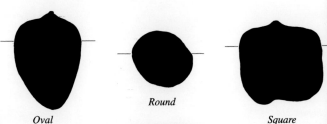

*Figure 215
Measuring the
width of scats of
different shapes*

Oval

Round

Square

PELLETS
Measure width at widest place

90

Use the shape and width to find the appropriate description below, then look at the illustrations on the pages listed for that description until you find an illustration matching the scat you are identifying.

Cylindrical scats with a width of

2 cm or more	see Plates 25–27	Pages 94–98
1 to 2 cm	see Plates 25–31, 48–51	Pages 94–106, 139–146
0.5 to 1 cm	see Plates 27–33, 53–55	Pages 98–110, 150–154
less than 0.5 cm	see Plates 33–41	Pages 110–126

Oval, round or square pellets with a width of

2 cm or more	see Plates 42–46	Pages 127–135
1 to 2 cm	see Plates 44–55, 30–32	Pages 131–154, 103–107
0.5 to 1 cm	see Plates 52–55, 35–41	Pages 147–154, 114–126
less than 0.5 cm	see Plates 33–41	Pages 110–126

Step 4   Use the distribution maps and habitat information for each species to help determine the species.

All photographs of scats in the Key were taken by Lindsay Addison of FOTOP, Belmont, Victoria.

Plate 25   Colour plate on page 94

1   Dingo *Canis lupus dingo*;
Dog (domestic and feral) *Canis familiaris*
All habitats with available drinking water, especially woodland
and grassland adjacent to forest (Text page 172)

2   Red Fox *Vulpes vulpes*
All habitats with available drinking water, especially woodland
and grassland adjacent to forest (Text page 172)

Plate 26   Colour plate on page 95

3   Short-beaked Echidna *Tachyglossus aculeatus*
Most habitats, from rainforest to desert (Text page 156)

4   Australian Fur-seal *Arctocephalus pusillus*
Waters of the continental shelf where there are rocky shores
(Text page 177)

Not illustrated:
Australian Sea-lion *Neophoca cinerea* (WA and SA coasts)
New Zealand Fur-seal Arctocephalus forsteri
        (Great Australian Bight; southern coasts of Tasmania)
Leopard Seal *Hydrurga leptonyx* (occasional sightings on
        south-eastern coasts, south-western WA and Tasmania)
Southern Elephant Seal *Mirounga leonina* (occasional
        sightings on south-eastern coasts and Tasmania)
Crab-eater Seal *Lobodon carcinophagus* (rare sightings on
        south-eastern coasts)

Plate 25 Maps on page 92

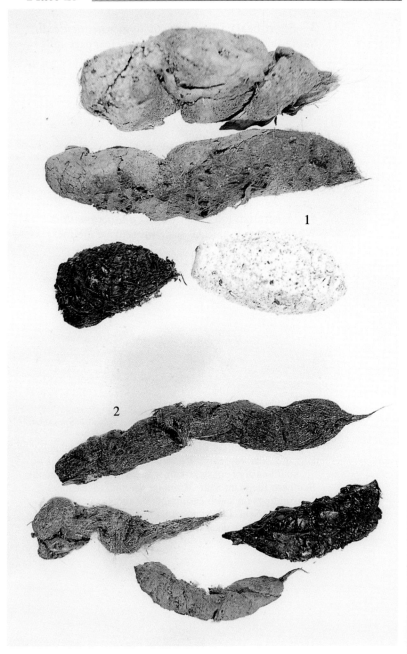

1

2

Plate 26   Maps on page 93

3

4

Plate 27   Colour plate on page 98

5   Tasmanian Devil *Sarcophilus harrisii*
Open forest and woodland; coastal scrub and agricultural areas
(Text page 167)

6   Spot-tailed Quoll *Dasyurus maculatus*
Open forest and rainforest; also dense coastal heathland
(Text page 167)

7   Cat *Felis catus*
All habitats (Text page 172)

Plate 28   Colour plate on page 99

8   Eastern Quoll *Dasyurus viverrinus*
Rainforest, open forest and scrub, heathland and moorland,
alpine areas and cultivated land (Text page 167)

9   Northern Quoll *Dasyurus hallucatus*
Woodland and scrub, particularly in rocky country
(Text page 167)

10   Western Quoll *Dasyurus geoffroii*
Most habitats, from wet forest to desert (Text page 167)

Plate 27 Maps on page 96

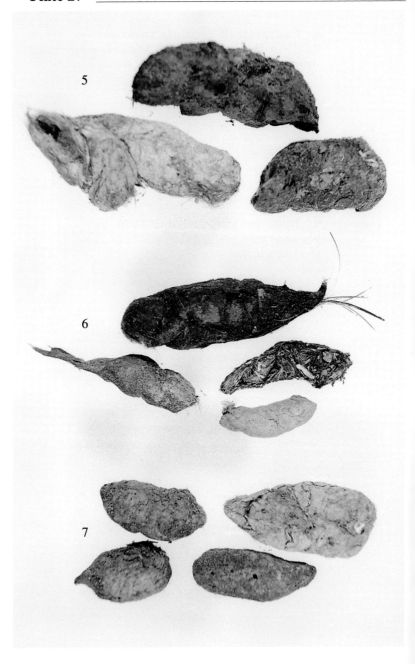

5

6

7

Plate 28   Maps on page 97

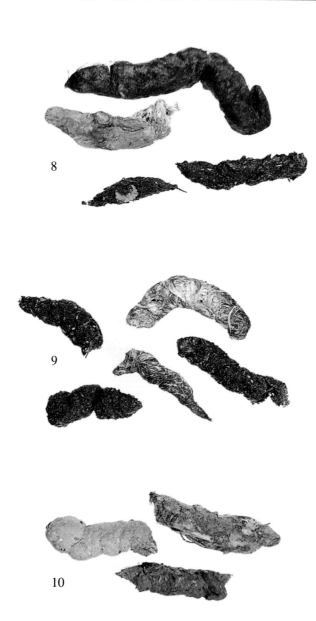

8

9

10

Plate 29   Colour plate on page 102

11 Long-nosed
Bandicoot
Perameles nasuta
Rainforest, wet and dry open
forest and woodland, with
grassy clearings
(Text page 166)

12   Northern Brown
Bandicoot
Isoodon macrourus
Open forest and woodland,
with dense understorey
(Text page 166)

13   Southern Brown
Bandicoot
Isoodon obesulus
Open forest and woodland,
with open understorey
(Text page 166)

14   Eastern Barred
Bandicoot
Perameles gunnii
Woodland and open grassland
(Text page 166)

Not illustrated:
Golden Bandicoot *Isoodon auratus* (coastal Kimberley region
    and Barrow Is, WA; and Marchinbar Is., NT).
Western Barred Bandicoot *Perameles bougainville* (islands
    off Shark Bay, WA)
Rufous Spiny Bandicoot *Echymipera rufescens* (Cape York,
    Qld)

Plate 30   Colour plate on page 103

15   Koala *Phascolarctos cinereus*
Open forest and woodland (Text page 165)

16   Mountain Brushtail Possum
Trichosurus caninus
Cool rainforest and tall, open forest (Text page 161)

17   Common Spotted Cuscus *Spilocuscus maculatus*
Lowland rainforest and adjacent mangroves (Text page 161)

Not illustrated:
Scaly-tailed Possum *Wyulda squamicaudata* (Kimberley region, WA)
Southern Common Cuscus *Phalanger intercastellanus* (Cape York, Qld)

Plate 29   Maps on page 100

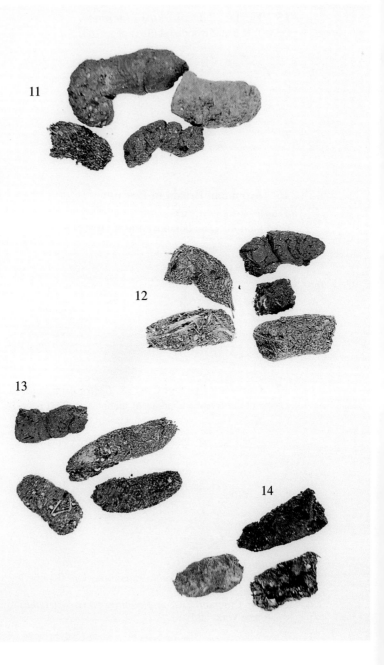

11

12

13

14

Plate 30   Maps on page 101

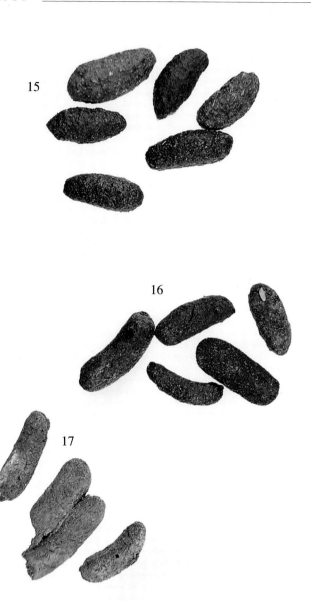

15

16

17

Plate 31   Colour plate on page 106

18   Common Brushtail Possum
Trichosurus vulpecula
Wet and dry open forest and woodland; towns and suburbs
(Text page 161)

The Common Brushtail Possum (**18**) and Northern Brushtail Possum (**19**) are now considered to be the same species – the Common Brushtail Possum *Trichosurus vulpecula*

19   Northern Brushtail Possum
Trichosurus arnhemensis
Open forest and woodland; suburban gardens
(Text page 161)

20   Rock Ringtail Possum
Petropseudes dahli
Rocky outcrops in open forest
(Text page 163)

21   Bilby
Macrotis lagotis
Arid and semi-arid woodland, hummock grassland
(Text page 166)

22   Numbat
Myrmecobius fasciatus
Dry eucalypt forest
(Text page 168)

Not illustrated:
Marsupial Mole *Notoryctes typhlops* (WA, SA and southern NT)

Plate 32   Colour plate on page 107

23   Common Ringtail Possum
Pseudocheirus peregrinus
Rainforest, open forest and woodland, especially thick scrub bordering swamps and streams; towns and suburbs (Text page 163)

24   Yellow-bellied Glider
Petaurus australis
Tall open forest and woodland, often in mountainous country (Text page 163)

25   Greater Glider
Petauroides volans
Tall open forest, woodland (Text page 163)

26   Platypus
Ornithorhynchus anatinus
Rivers and freshwater lakes where burrows can be dug on the banks (Text page 156)

27   Lemuroid Ringtail Possum
Hemibelideus lemuroides
Upland rainforest (Text page 163)

28   Herbet River Ringtail Possum
Pseudochirulus herbertensis
Rainforest (Text page 163)

Not illustrated: Green Ringtail Possum   *Pseudochirops archeri* (northern Qld)
Daintree River Ringtail Possum   *Pseudochirulus cinereus* (northern Qld)
Western Ringtail Possum   *Pseudocheirus occidentalis* (south-west WA)

105

Plate 31   Maps on page 104

18

19

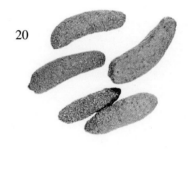

20

21

22

Plate 32 Maps on page 105

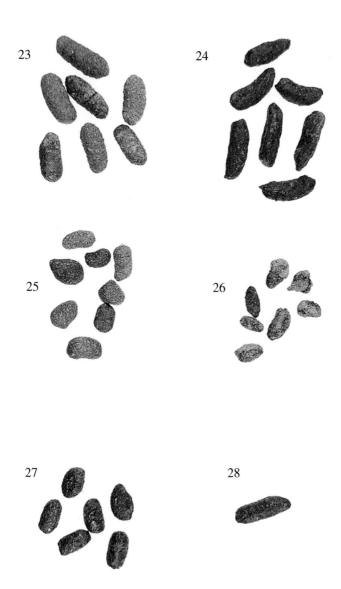

23

24

25

26

27

28

Plate 33   Colour plate on page 110

29 Water-rat *Hydromys chrysogaster*
Creeks, rivers and lakes with undergrowth on the banks, in most habitats; also offshore islands and estuaries
(Text page 170)

30 Brush-tailed Phascogale
Phascogale tapoatafa
Dry open forest and woodland, particularly on ridges and rocky slopes (Text page 167)

31 Kowari
Dasyuroides byrnei
Stony desert
(Text page 167)

32 Dusky Antechinus
Antechinus swainsonii
Rainforest, wet forest with dense undergrowth; also in coastal and alpine heathland
(Text page 167)

33 Mulgara
Dasycercus cristicauda
Sandy desert
(Text page 167)

Not illustrated:
False Water-rat *Xeromys myoides* (Arnhem Land, NT; south-eastern Qld)

Plate 34 Colour plate on page 111

34 Swamp Rat
Rattus lutreolus
Wet heath and sedgeland,
dense grassland
(Text page 169)

35 Bush Rat
Rattus fuscipes
Rainforest, open forest with
good ground cover, wood-
land, coastal scrub
(Text page 169)

36 Broad-toothed Rat
Mastacomys fuscus
Open forest with dense
undergrowth; wet alpine and
subalpine grassland and
dense sedgeland
(Text page 169)

37 Black Rat
Rattus rattus
Most well-watered areas,
especially in or near human
habitation
(Text page 169)

38 Brown Rat
Rattus norvegicus
In or near human habitation,
especially in coastal towns
(Text page 169)

39 Canefield Rat
Rattus sordidus
Canefields, tropical grassland
and open forest
(Text page 169)

Plate 33   Maps on page 108

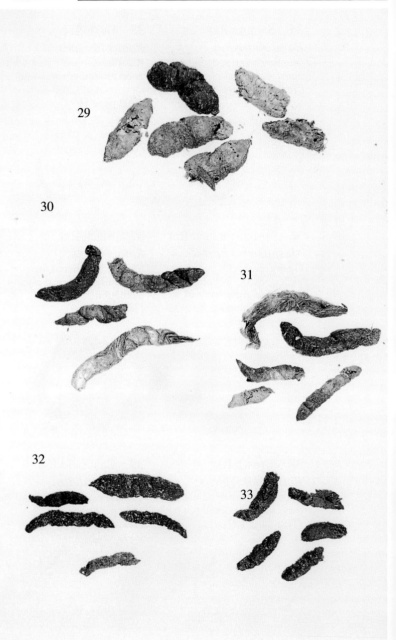

29

30

31

32

33

Plate 34   Maps on page 109

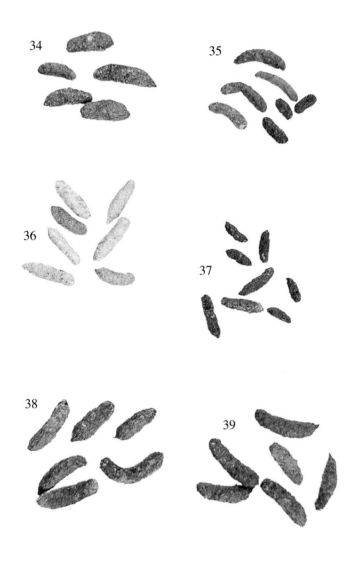

34

35

36

37

38

39

Plate 35   Colour plate on page 114

40   Pale Field-rat
Rattus tunneyi
Tall grassland, usually near water (Text page 169)

41   Giant White-tailed Rat
Uromys caudimaculatus
Rainforest and wet open forest (Text page 169)

42   Dusky Rat
Rattus colletti
Alluvial flood plains bordered by grasses and sedge (Text page 169)

43   Grassland Melomys
Melomys burtoni
Coastal tall grassland (Text page 169)

44   Fawn-footed Melomys
Melomys cervinipes
Rainforest and wet open forest (Text page 169)

45   Masked Uromys
Uromys hadrourus
Rainforest (Text page 169)

Not illustrated:
Cape York Rat *Rattus leucopus* (Cape York, Qld)
Long-haired Rat *Rattus villosissimus* (north-western and south-western Qld; Simpson Desert; Barkly Tableland)
Cape York Melomys *Melomys capensis* (Cape York, Qld)

Plate 36   Colour plate on page 115

46   Musky Rat-kangaroo
Hypsiprymnodon moschatus
Rainforest (Text page 160)

47   Sugar Glider
Petaurus breviceps
Wet and dry open forest and woodland, usually with acacia in the understorey (Text page 163)

48   Squirrel Glider
Petaurus norfolcensis
Dry open forest and woodland (Text page 163)

49   Leadbeater's Possum
Gymnobelideus leadbeateri
Mostly tall open forest, mainly confined to Mountain Ash forest (Text page 163)

50   Yellow-footed Antechinus *Antechinus flavipes*
Rainforest, wet and dry open forest (especially box and stringybark) and woodland (Text page 167)

51   Brown Antechinus
Antechinus stuartii
Wet and dry open forest, particularly with dense groundcover (Text page 167)

See page 153 for similar species.

Plate 35   Maps on page 112

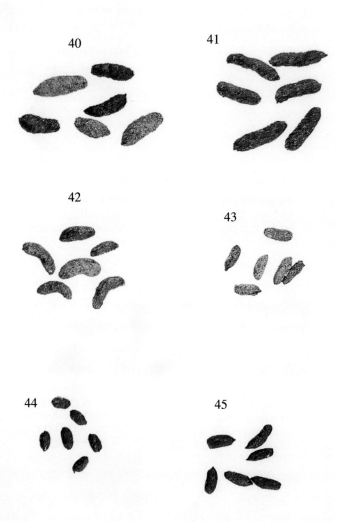

40

41

42

43

44

45

Plate 36   Maps on page 113

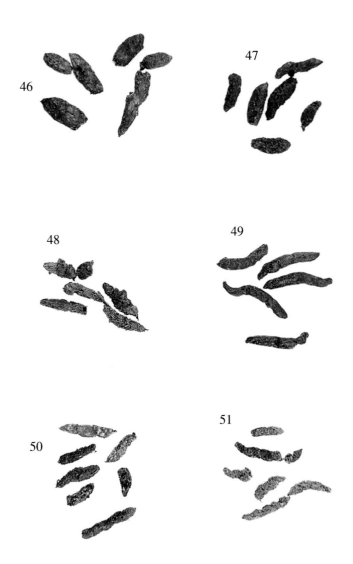

46

47

48

49

50

51

Plate 37   Colour plate on page 118

52   Red-tailed Phascogale
Phascogale calura
Dry mature eucalypt forest
with continuous canopy
(Text page 168)

53   Common Dunnart
Sminthopsis murina
Dry open forest, woodland
and mallee heath
(Text page 168)

54   Fat-tailed Dunnart
Sminthopsis crassicaudata
Woodland, shrubland, tussock
grassland, gibber plain and
agricultural land
(Text page 168)

55   White-footed Dunnart
Sminthopsis leucopus
Open forest and woodland with
heathy understorey; coastal
heaths, sedgeland and tussock
grassland (Text page 168)

56   Red-cheeked Dunnart
Sminthopsis virginiae
Open savanna woodland
(Text page 168)

57   Mallee Ningaui
Ningaui yvonneae
Sandy country with spinifex
hummock grasslands and
mallee scrub (Text page 168)

Not illustrated:
Dunnarts *Sminthopsis* species (about 13 other species around
    Australia)
Kultarr *Antechinomys laniger* (central-western WA; central
    Australia)

Plate 38 Colour plate on page 119

58 Common Rock-rat
Zyzomys argurus
Rocky outcrops in open
forest, woodland or grassland
(Text page 169)

59 Brush-tailed Rabbit-rat
Conilurus penicillatus
Open eucalypt woodland,
monsoon forest and pandanus
scrub (Text page 169)

60 Heath Mouse
Pseudomys shortridgei
Heath and open forest with
heathy understorey, espe-
cially in regrowth after fire
(Text page 169)

61 Smoky Mouse
Pseudomys fumeus
Subalpine open forest and
woodland with heathy
understorey; coastal heath
(Text page 169)

62 Mountain Pygmy-
possum
Burramys parvus
Alpine scrub and Snow Gum
woodland with boulders,
above 1400 metres
(Text page 164)

63 Eastern
Pygmy-possum
Cercartetus nanus
Rainforest; wet open forest,
shrubland and heath;
subalpine woodland
(Text page 164)

See page 153 for similar species

117

Plate 37   Maps on pag 116

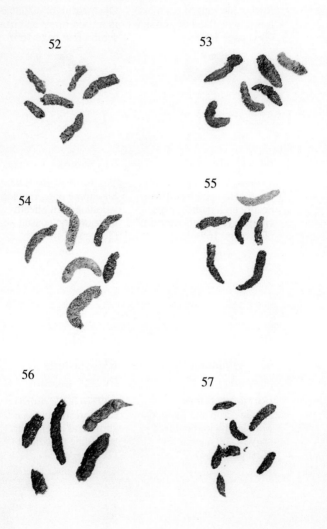

52

53

54

55

56

57

Plate 38 Maps on page 117

58

59

60

61

62

63

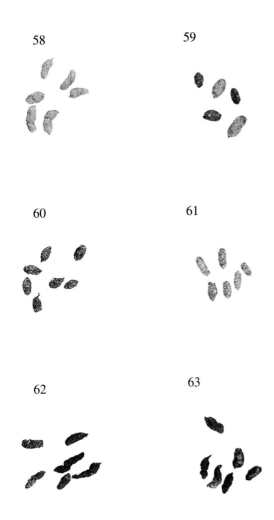

Plate 39   Colour plate on page 122

64 Silky Mouse
Pseudomys apodemoides
Dry heath, especially
associated with Desert
Banksia (Text page 169)

65 Long-tailed Mouse
Pseudomys higginsi
Rainforest and wet forest;
also in subalpine areas
(Text page 169)

66 Plains Mouse
Pseudomys australis
Gibber plains
(Text page 169)

67 Ash-grey Mouse
Pseudomys albocinereus
Woodland and heath on sandy
soil (Text page 169)

68 Common Planigale
Planigale maculata
Rainforest, open forest,
woodland and marsh
(Text page 168)

69 Wongai Ningaui
Ningaui ridei
Dunes in sandy desert where
there are spinifex hummocks
(Text page 168)

Not illustrated: Giles' Planigale *Planigale gilesi* (western and central NSW)
Long-tailed Planigale *Planigale ingrami* (Kimberley region,
    NT; Gulf Country, NT; Qld)
Narrow-nosed Planigale *Planigale tenuirostris* (western and
    central NSW)
Pilbara Ningaui *Ningaui timealeyi* (Pilbara region, WA)

Plate 40   Colour plate on page 123

70 Mitchell's Hopping-mouse *Notomys mitchelli*
Mallee heathland
(Text page 170)

71 Spinifex Hopping-mouse *Notomys alexis*
Sandy desert and spinifex grassland (Text page 170)

72 Dusky Hopping-mouse *Notomys fuscus*
Desert sand dunes
(Text page 170)

73 Sandy Inland Mouse *Pseudomys hermannsburgensis*
Mulga scrub, sandy desert and gibber flats, hummock grasslands (Text page 169)

74 House Mouse *Mus musculus*
Most common near human habitation; also in most habitats except rainforest, especially after fire (Text page 169)

Not illustrated:
About 11 other species of Pseudomys around Australia
Prehensile-tailed Rat *Pogonomys mollipilosus* (northern Qld)
Fawn Hopping-mouse *Notomys cervinus* (Lake Eyre basin; south-western Qld)
Northern Hopping-mouse *Notomys aquilo* (Cape York, Qld; Groote Eylandt, NT)

121

Plate 39 Maps on page 120

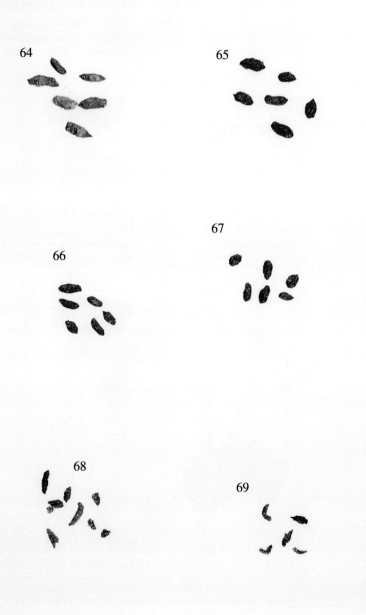

64

65

67

66

68

69

Plate 40 Maps on page 121

70

71

72

73

74

Plate 41 Colour plate on page 126

75 New Holland Mouse
Pseudomys novaehllandiae
Low heathland on sandy soil, especially in regrowth after fire (Text page 169)

76 Feathertail Glider
Acrobates pygmaeus
Open forest and woodland (Text page 164)

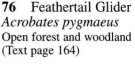

77 Little Pygmy-possum
Cercartetus lepidus
Mallee heath; open forest and heathland in Tasmania (Text page 164)

78 Western Pygmy-possum
Cercartetus concinnus
Dry open forest with dense understorey, woodland and mallee heath (Text page 164)

79 Eastern Horseshoe-bat
Rhinolopus megaphyllus
Wet and dry open forest (Text page 177)

80 Common Bent-wing Bat
Miniopterus schreibersii
Rainforest, wet and dry open forest (Text page 177)

Not illustrated:
Long-tailed Pygmy-possum *Cercartetus caudatus* (northern Qld)
Honey Possum *Tarsipes rostratus* (south-western WA)
microbats: about 60 other species found around Australia

Plate 42   Colour plate on page 127

81   Feral Pig *Sus scrofa*
Most habitats with available water and dense vegetation
(Text page 176)

82   One-humped Camel *Camelus dromedarius*
Sandy desert (Text page 176)

Plate 41 Maps on page 124

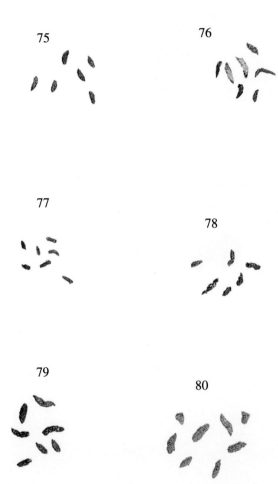

75

76

77

78

79

80

Plate 42   Maps on page 125

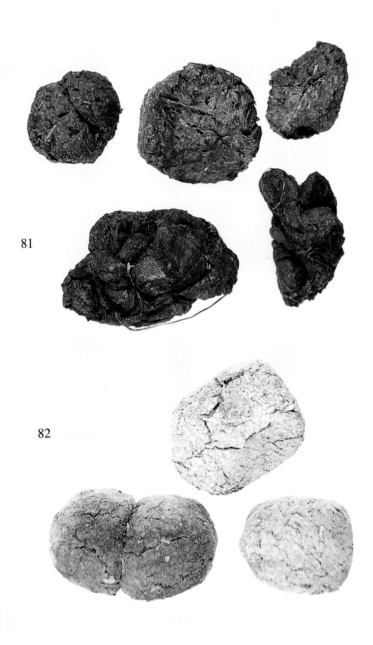

81

82

Plate 43   Colour plate on page 130

83   Common Wombat   *Vombatus ursinus*
Wet and dry open forest with dense undergrowth and grassy areas; also alpine woodland and sandy heathland (Text page 165)

85   Southern Hairy-nosed Wombat
Lasiorhinus latifrons
Semi-arid woodland and shrubland (Text page 165)

85   Northern Hairy-nosed Wombat
Lasiorhinus krefftii
Semi-arid woodland and grassland (Text page 165)

Plate 44   Colour plate on page 131

86   Eastern Grey Kangaroo *Macropus giganteus*
Open forest, woodland and shrubland with grassy clearings
(Text page 157)

87   Western Grey Kangaroo *Macropus fuliginosus*
Dry open forest, woodland and mallee with grassy areas
(Text page 157)

88   Red Kangaroo *Macropus rufus*
Grassland and saltbush plains with timber or scrub belts
(Text page 157)

Plate 43 Maps on page 128

83

84

85

Plate 44   Map on page 129

86

87

88

Plate 45   Colour plate on page 134

89   Swamp Wallaby *Wallabia bicolor*
Rainforest, open forest and woodland with dense understorey
(Text page 157)

90   Red-necked Wallaby *Macropus rufogriseus*
Wet and dry open forest and woodland with grassy areas
(Text page 157)

91   Whiptail Wallaby *Macropus parryi*
Open forest with adjacent grassy areas (Text page 157)

Plate 46   Colour plate on page 135

92 Common Wallaroo *Macropus robustus*
Varied, from wet open forest to arid grassland, usually in rocky or stony hill country (Text page 157)

93 Antilopine Wallaroo *Macropus antilopinus*
Open eucalypt woodland with grassy understorey (Text page 157)

94 Agile Wallaby *Macropus agilis*
Open forest and woodland with adjacent grassy areas; coastal sand dunes (Text page 157)

Not illustrated:
Black Wallaroo *Macropus bernardus* (Arnhem Land, NT) (Text page 157)

Plate 45 Maps on page 132

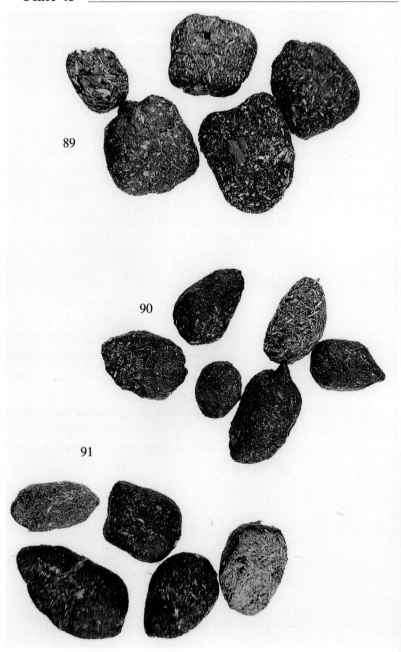

89

90

91

Plate 46   Maps on page 133

92

93

94

Plate 47   Colour plate on page 138

95   Black-striped
Wallaby
Macropus dorsalis
Open forest, brigalow scrub
and woodland with dense
understorey (Text page 157)

96   Parma Wallaby
Macropus parma
Wet mountain forest with
dense understorey and
adjacent grassy areas
(Text page 157)

97   Tammar Wallaby
Macropus eugenii
Open forest, coastal scrub
and woodland with adjacent
grassy areas (Text page 157)

98   Quokka
Setonix brachyurus
Open forest, woodland and
semi-arid heath
(Text page 159)

Not illustrated:
Western Brush Wallaby *Macropus irma* (south-western WA)

Plate 48   Colour plate on page 139

99   Rufous Bettong
Aepyprymnus rufescens
Open forest with dense
understorey and grassy areas
(Text page 160)

100   Northern Nailtail Wallaby
Onychogalea unguifera
Open woodland with grassy
understorey; grassland
(Text page 159)

101   Bridled Nailtail Wallaby
Onychogalea fraenata
Woodland and scrub with
grassy areas (Text page 159)

102   Lumholtz's Tree-kangaroo
Dendrolagus lumholtzi
Upland rainforest
(Text page 159)

Not illustrated:
Bennett's Tree-kangaroo *Dendrolagus bennettianus* (northern
    Qld)

Plate 47   Maps on page 136

95

96

97

98

Plate 48   Maps on page 137

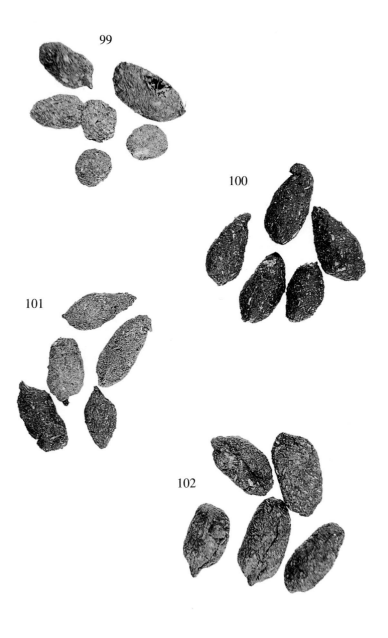

99

100

101

102

Plate 49   Colour plate on page 142

103   Tasmanian Pademelon   *Thylogale billardierii*
Rainforest and wet and dry eucalypt forest with dense understorey, particularly with adjacent cleared areas (Text page 159)

104   Red-necked Pademelon   *Thylogale thetis*
Rainforest and wet eucalypt forest with adjacent grassy areas or pasture (Text page 159)

105   Red-legged Pademelon   *Thylogale stigmatica*
Rainforest and wet eucalypt forest (Text pages 159)

Plate 50   Colour plate on page 143

106   Brush-tailed Rock-wallaby   *Petrogale penicillata*
Steep rocky cliffs and ridges with nearby grassy areas, usually
in open forest (Text page 159)

107   Yellow-footed Rock-wallaby
Petrogale xanthopus
Rocky cliffs and outcrops in arid, open woodland and
acacia scrub (Text page 159)

108   Unadorned Rock-wallaby   *Petrogale inornata*
Steep slopes of rainforest, open forest and woodland with some
rocky outcrops (Text page 159)

Not illustrated:
Short-eared Rock-wallaby *Petrogale brachyotis* (Kimberley
     Region, WA; Arnhem Land, NT)
Monjon *Petrogale burbidgei* (Kimberley Region, WA)
Black-footed Rock-wallaby *Petrogale lateralis*
     (widely scattered in ranges of WA and NT)
Rothschild's Rock-wallaby *Petrogale rothschildi*
     (Pilbara Region, and islands in Dampier Archipelago, WA)

Plate 49 Maps on page 140

103

104

105

Plate 50   Maps on page 141

106

107

108

Plate 51   Colour plate on page 146

109   Proserpine Rock-wallaby

Petrogale persephone
Rocky outcrops in patches of rainforest, surrounded by woodland with grassy understorey (Text page 159)

110   Purple-necked Rock-wallaby

Petrogale purpureicollis
Granit rock-piles in arid and semi-arid scrub country
(Text page 159)

111   Allied Rock-wallaby   *Petrogale assimilis*
Steep slopes of rainforest, wet open forest and woodland
(Text page 159)

Not illustrated: 5 other species of Rock Wallaby found in
north-east Qld
Cape York Rock-wallaby *Petrogale coenensis*
Godman's Rock-wallaby *Petrogale godmani*
Herbert's Rock-wallaby *Petrogale herberti*
Mareeba Rock-wallaby *Petrogale mareeba*
Sharman's Rock-wallaby *Petrogale sharmani*

Plate 52   Colour plate on page 147

112   Spectacled Hare-wallaby

Lagorchestes conspicillatus
Open forest and woodland, shrubland and tussock grassland
(Text page 159)

113   Rufous Hare-wallaby (Mala)

Lagorchestes hirsutus
Woodland, scrubland and grassland, especially during regrowth
after fire (Text page 159)

114   Nabarlek   *Petrogale concinna*
Rocky areas in tropical grassland (Tect page 159)

Not illustrated:
Banded Hare-wallaby *Lagostrophus fasciatus* (islands off
    Shark Bay, WA)

Plate 51   Map on page 144

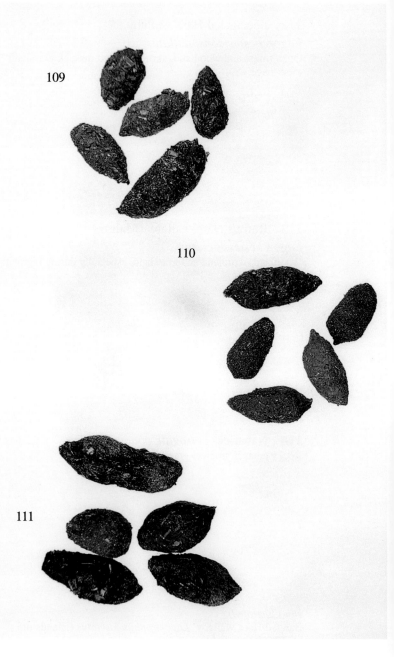

109

110

111

Plate 52 Maps on page 145

112

113

114

Plate 53   Colour plate on page 150

115   Tasmanian Bettong
Bettongia gaimardi
Dry open forest and wood-
land with open, grassy
understorey (Text page 160)

116   Brush-tailed
Bettong (Woylie)
Bettongia penicillata
Open forest and woodland
with low shrubs or tussocks
(Text page 160)

117   Long-footed Potoroo   *Potorous longipes*
Wet forest with dense understorey (Text page 160)

118   Long-nosed Potoroo
Potorous tridactylus
Cool rainforest, wet open
forest with dense understorey,
dense coastal heathland
(Text page 160)

119   Northern Bettong
Bettongia tropica
Open forest and woodland
with low shrubs or tussocks,
on edge of rainforest
(Text page 160)

Not illustrated:
Burrowing Bettong *Bettongia leseur* (islands off WA coast)
Gilbert's Potoroo *Potorous gilberti* (south-west WA)

Plate 54   Colour plate on page 151

120   Red Deer
Cervus elaphus
Open forest and woodland
(Text page 174)

121   Sambar
Cervus unicolor
Open forest with dense
understorey
(Text page 174)

122   Fallow Deer
Dama dama
Open forest and woodland
(Text page 174)

123   Hog Deer
Axis porcinus
Coastal scrub, swampy
woodland and river flats
(Text page 174)

Not illustrated:
Chital *Axis axis* (northern Qld)
Water Buffalo *Bubalus bubalis* (northern NT)
Banteng *Bos javanicus* (Coburg Peninsula, NT)
Horse *Equus caballus* (northern WA, NT, Qld, western NSW,
    north-eastern SA, alpine Victoria and NSW)

Illustrated elswhere:
Donkey *Equus asinus* (WA, NT, south-western Qld, SA:
    page 174)
Domestic Cattle *Bos taurus* (small populations of feral
    animals in many areas; managed herds: page 176)

Plate 53   Maps on page 148

115

116

117

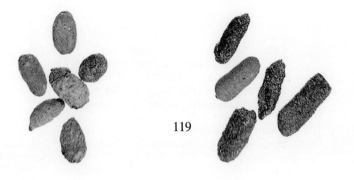

119

Plate 54   Maps on page 149

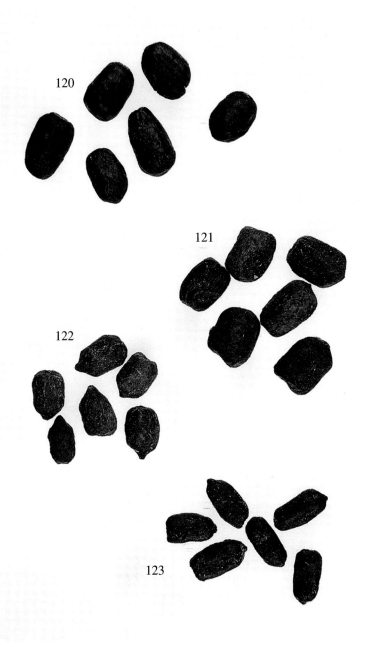

120

121

122

123

Plate 55 Colour plate on page 154

124 Rusa Deer
Cervus timoriensis
Open forest and grassland
(Text page 174)

125 Feral Goat
Capra hircus
Most habitats except rainforest,
wetlands and desert; favours
rocky areas, dense cover and
available water
(Text page 174)

126 Sheep
Ovis aries
Small populations of feral
sheep live in a variety of
habitats; domestic herds
mostly in grasslands
(Text page 174)

127 Brown Hare
Lepus capensis
Open grassland, woodland
(Text page 171)

128 European Rabbit *Oryctolagus cuniculus*
Most habitats except rainforest, especially pasture and grazing land with some ground cover (Text page 171)

Plate 36 Continued from page 113

Not illustrated:
Mahogany Glider *Petaurus gracilis* (northern Qld)
Striped Possum *Dactylopsila trivirgata* (northern Qld)
Swamp Antechinus *Antechinus minimus* (Tasmania; south-eastern SA; southern coastal Victoria)
Fawn Antechinus *Antechinus bellus* (arnhem Land, NT)
Cinnamon Antechinus *Antechinus leo* (northern Qld)
Southern Dibbler *Parantechinus apicalis* (southern WA)
Northern Dibbler *Parantechinus bilarni* (Arnhem Land, NT)
Carpentarian Pseudantechinus *Pseudantechinus mimulus* (north-west Qld, north-east NT)
Fat-tailed Pseudantechinus *Pseudantechinus macdonnellensis* (MacDonnell Ranges, Tanami Desert, NT; Pilbara region, WA)
Ningbing Pseudantechinus *Pseudaentechinus ningbing* (Kimberley region, NT)
Tan Pseudantechinus *Pseudantechinus roryi* (Pilbara region, WA)
Woolley's Pseudantechinus *Pseudantechinus woolleyae* (Port Hedland area, WA)
Little Red Kaluta *Dasykaluta rosamondae* (Pilbara region, WA)

Plate 38 Continued from page 117

Not illustrated:
Black-footed Tree-rat *Mesembriomys gouldii* (Kimberley region, WA; northern NT; northern Qld)
Golden-backed Tree-rat *Mesembriomys macrurus* (Kimberley region, WA; northern Qld)
Large Rock-rat *Zyzomys woodwardi* (Kimberley region, WA; Alligator Rivers region, NT)
Central Rock-rat *Zyzomys pedunculatus* (central Australia)
Several rare species, including: Forrest's Mouse *Leggadina forresti* (North-west WA and central Australia), Lakeland Downs Mouse *Leggadina lakedownensis* (northern Qld), Greater Stick-nest Rat *Leporillus conditor* (Franklin Is. in Nuyts Archipelago, SA, recently released at several mainland and island sites in WA and SA), Arnhem Land Rock-rat *Zyzomys maini* (Arnhem Land plateau, NT), Carpentarian Rock-rat *Zyzomys palatalis* (Gulf Country, NT).

Plate 55 Maps on page 152

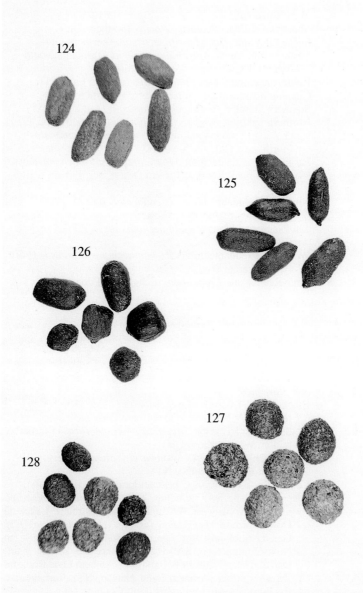

124

125

126

127

128

Plate 56 Text on page 178

OWL PELLETS

Barn Owl

Sooty Owl

Powerful Owl

Masked Owl

Guide to Scats

Monotremes

Echidna

Echidna scats are long cylinders, up to 2 cm in diameter. They are smooth and pliable when fresh, and sometimes have a film of mucus. When dry, small insect particles are usually visible on the surface. They become very friable and are often found broken into two or more pieces.

The scats contain many fragments of insects, mainly ants and termites but also sometimes beetles and the papery skins of beetle larvae and earthworms, as well as soil which has been ingested with the insects. The colour of the scats varies with the colour of the soil in the area where the echidna has been feeding. They have a strong, earthy smell when fresh, but dry scats have little or no smell.

Echidna scats can be found anywhere. They are often found on rock piles, and also where the echidna has been feeding; for example, near termite mounds and meat-ant mounds.

Key to Scats No 3, Plate 26, page 95

Platypus

Platypus scats are uneven, mud-like pellets. They contain very fine particles, mostly of silt from the water in which the Platypus has been feeding. They have little or no odour.

Scats are usually deposited in the water, but occasionally they can be found near a burrow entrance (Plate 60, page 197).

Key to Scats No 26, Plate 32, page 107

Marsupials

KANGAROOS AND THEIR RELATIVES

Kangaroos, Wallaroos and Large Wallabies

Scats of these animals are usually deposited as separate pellets in groups of four to eight. The pellets are oval, round or square, and in a group of scats from one animal there may be pellets of two or more different shapes. In most species, a coating of mucus forms a shiny surface when the pellets are fresh. This dries to a black or dark brown coating (Figure 216).

When pellets are broken, the plant material can be seen. The texture of this is often a useful clue in identification, as some macropods eat softer grasses than others, and some browse on bushes and trees. For example, Eastern Grey Kangaroo scats are green or green-brown when broken and the material is uniform and close-textured, as this species eats mainly the softer grasses; Swamp Wallaby scats are usually brown when broken and have a coarser, less uniform texture, as some woody plants are eaten as well as grasses. Broken pellets of the larger macropods are illustrated in the Key to Scats.

Fresh scats have a strong odour, similar to plant humus. Dry scats have little or no odour.

Pellets are often deposited in a clump and are sometimes joined in a string. In spring and summer, when the fresh growth of grass is eaten, the pellets may be unformed and the scats extruded as unsegmented cylinders (Figure 217).

These cylindrical scats can be distinguished from the similarly shaped scats of some large carnivores by their lack of strong, acrid odour and their contents of plant material.

The scats of all these animals are found around their feeding areas and sleeping places, and on their paths through the undergrowth.

Key to Scats Nos 86–97, Plates 44, 45, 46, 47, pages 131, 134, 135, 138

*Figure 216
A fresh deposit
of Eastern Grey
Kangaroo scats*

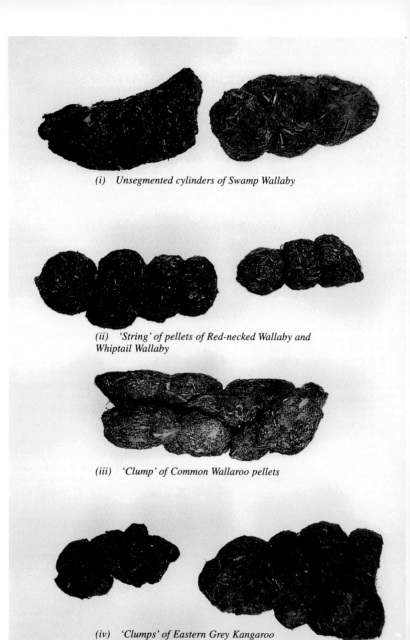

(i)   *Unsegmented cylinders of Swamp Wallaby*

(ii) *'String' of pellets of Red-necked Wallaby and Whiptail Wallaby*

(iii) *'Clump' of Common Wallaroo pellets*

(iv) *'Clumps' of Eastern Grey Kangaroo pellets, young and adult*

Figure 217   Cylinders, 'strings' and 'clumps' of macropod scats

Pademelons, Rock-wallabies, Tree-kangaroos, Nailtail Wallabies, Hare-wallabies and Quokka

These smaller macropods also deposit their scats as pellets in groups, usually of four to eight at a time. Some pellets are oval or round, but others are more cylindrical. Hare-wallaby scats are slightly flattened. Most scats have a coating of mucus when fresh (more noticeable in the species from the semi-arid areas) which dries to a black or brown glaze.

When broken, the scats show fairly coarse fragments of plant material, usually brown. The material in the scats of some species is more tightly packed than in others — some rock-wallaby scats, for example, are loosely packed and have a chaff-like appearance. In dry caves and ledges the scats of rock-wallabies take a long time to decompose, and large accumulations may result.

The scats have a strong smell, similar to plant humus, when fresh, but little odour when dry. They are found around feeding areas and sleeping places, or scattered on paths through the undergrowth. Tree-kangaroos sometimes deposit their scats on rocks.

Key to Scats Nos 98, 100–114, Plates 47–52, pages 138, 139, 142, 143, 146, 147

Bettongs, Potoroos and Musky Rat-kangaroo

Scats of these animals are much harder to find than those of the larger macropods, as most of them are small and disappear in the dense undergrowth, where they are likely to be eaten by dung beetles. They are generally dark brown or black, shiny when fresh with a smooth, mud-like coating that is also smooth when dry.

The Rufous Bettong's scats contain fairly coarse plant material, but as the other bettongs and potoroos feed on fungi — mostly the fruiting bodies of underground fungi — and soft-bodied insect larvae and other invertebrates, their scats are made up of fine particles of soil and fungal spores. Finding the fruiting bodies (sporocarps) of the fungi on the ground may help locate the scats (see Plate 66, page 199).

Some fibrous material from bulbs and seeds may also be found in the scats, especially in winter and spring, as these foods are eaten when fewer fungi are available. Musky Rat-kangaroo scats contain fine particles of fibrous material from the fruits of rainforest trees, and some insect material.

Fresh scats of the Rufous Bettong have a smell similar to plant humus. Scats of other bettongs and potoroos have a strong, almost meaty odour, which often lingers even when the scats are old and dry.

The Rufous Bettong leaves large areas of disturbed ground where it has foraged for roots and tubers, and its scats may be found near these diggings. Bettong and potoroo scats are deposited near their feeding sites, which are marked by small conical holes where the animal has been digging (Plates 63, 64, 65, page 199). The Musky Rat-kangaroo's scats are rarely found, as it lives in rainforest.

Key to Scats Nos 46, 99, 115–119, Plates 36, 48, 53, pages 113, 139, 150

POSSUMS

Brushtail possums, scaly-tailed Possum and Cuscuses

Cylindrical pellets are produced by all these species. The large scats of the Mountain Brushtail Possum and the Common Spotted Cuscus consist of fairly coarse fragments of plant material. Common Brushtail Possums also eat much plant material but occasionally eat insects, especially large moths when these are available, and fragments of insects are sometime found in their scats. They have been seen taking eggs and nestlings from birds' nests, and will eat meat scraps and other garbage in cities and suburbs, so remains of these foods are present in some scats They also lick sap from eucalypts at feeding sites made by gliders, and their scats are then dark and contain fine powdery material.

The scats of the Common Brushtail Possum are probably the most variable in size and colour of any species. They vary in colour from red-brown to black, and in shape from fat, round cylinders to narrow, rat-like scats. They are usually deposited as groups of single pellets, but may also be grouped in clumps or as strings, sometimes connected by hair or by plant material (Figure 218). Southern Common Cuscus and Scaly-tailed Possum scats were not available for examination.

Most fresh brushtail scats have a mild musty smell, but fresh scats of the Mountain Brushtail Possum have a rich, pungent odour. Sometimes brushtail scats have a strong smell of eucalyptus oil.

Common Brushtail Possums often live near human habitation, so their scats are frequently found on verandahs and driveways, near fruit trees, on fences and in similar places. If there is a pet dog, possum scats are often found near the dog's food bowl. In forests, their scats are found near the bases of their den trees and where they have been feeding. Mountain Brushtail Possum scats are often found on fallen logs and near regularly used trees, but cuscus scats are difficult to find as these animals live in rainforest.

Key to Scats Nos 16–19, Plates 30, 31, pages 103, 106

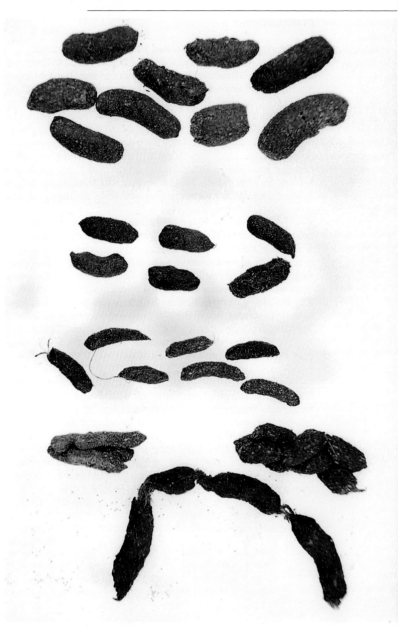

Figure 218   Assorted scats of the Common Brushtail Possum — large, medium and small — 'clumps' and a 'string'

Ringtail Possums

Cylindrical scats, generally smaller than most brushtail possum scats, are deposited in groups of about three to six at a time, but sometimes singly. They have a granulated surface and both ends are usually rounded. The scats of the Greater Glider are similar in some ways, but ringtail scats are more regularly cylindrical.

Common Ringtail Possum scats are often red-brown, but the colour varies with the diet: they may also be dark green, brown or grey. The banana-shaped scats of the Rock Ringtail also vary in colour. Few scats of the rainforest ringtails have been found.

Ringtails feed mainly on leaves, supplemented by blossoms and fruit, and as their teeth are able to chop up this food very finely the scats contain fine particles of plant material. Fresh scats have a mild smell of eucalyptus oil, but older scats have no odour.

Scats of the Common Ringtail Possum are found near the bases of the trees where these animals have their nests and where they have been feeding (see Plates 70, 71, page 201). As they often live near human habitation, their scats are also found in gardens and on verandahs and paths. Rock Ringtails often leave their scats on rocks. Scats of the rainforest ringtails are rarely found because they disappear in the undergrowth.

Key to Scats Nos 20, 23, 27, 28, Plates 31, 32, pages 106, 107

Gliders, Striped Possum and Leadbeater's Possum

There is considerable variety among the scats of this group of possums. The scats of the Sugar Glider, Squirrel Glider and Leadbeater's Possum are long, narrow cylinders, usually friable, dark and shiny on the surface. They are sometimes paler when the animal has been feeding mainly on wattle gum. These species' diet consists of wattle gum and eucalypt sap, with some insects, manna and honeydew, so their scats contain fine powdery particles. They have a strong, sweet smell when fresh but little odour when dry. Mahogany Glider scats are said to be similar to Squirrel Glider scats.

Yellow-bellied Glider scats are also cylindrical, and are dark and shiny. They contain fine brown powdery material, the result of the sap, nectar and pollen in the diet. Small fragments of insects, such as cicadas, are occasionally present. Fresh scats have a strong, sweet odour.

Greater Gliders eat eucalypt leaves and, like the ringtails, they masticate them very finely, so the scats contain very fine leaf fragments. Although generally small and cylindrical, their shape is not as regular as ringtail scats, and this diversity of shape is a good diagnostic feature of Greater Glider scats. They are usually red-brown or brown and have little or no odour. They are found in large numbers under the den trees, as these gliders defecate soon after emerging from the den.

The Striped Possum eats mainly wood-boring grubs and other insects, and some leaves and fruit. The scats, which probably contain fine insect remains, disappear in the rainforest leaf litter.

Scats of the gliders are found at the bases of trees and are sometimes caught in the rough bark near sap-feeding sites (see Plates 73, 74, 75, 77, pages 201, 202, 203).

Key to Scats Nos 24, 25, 47, 48, 49, Plates 32, 36, pages 107, 115

Pygmy-possums, Feathertail Glider and Honey Possum

Scats of these species are small and similar in shape to those of mice. They consist of very fine material — the residue of insects, nectar and pollen — and, in the Mountain Pygmy-possum's scats, fragments of seeds and berries as well as insects.

Fresh scats of pygmy-possums have a strong, sweet smell, but dry scats have little odour.

Most pygmy-possum scats, and also those of the Feathertail Glider and Honey Possum, are not easy to find because they are so small that they disappear in the leaf litter, but Mountain Pygmy-possum scats are occasionally found scattered on boulders.

Key to Scats Nos 62, 63, 76, 77, 78, Plates 38, 41, pages 119, 126

Figure 219 Scats of the Common Wombat deposited on a 'prominent' mushroom

KOALA AND WOMBATS

Koala

The hard, firmly packed scats of the Koala — long, oval or cylindrical and with a slightly ridged surface — are easy to identify. Their colour varies, but most are brown or red-brown. Blue-green, grey-green and yellow-brown Koala scats are also found at times.

Koala scats contain fairly coarse fragments of leaf cuticles, chiefly of the eucalypts that are the main food. Some of the eucalypts favoured by Koalas are the Manna Gum (*Eucalyptus viminalis*), Forest Red Gum (*E. tereticornis*), River Red Gum (*E. camaldulensis*), Grey Gum (*E. cypellocarpa*), Swamp Gum (*E. ovata*) and Long-leaf Box (*E. goniocalyx*), but they also feed in a number of other eucalypts and in other species of trees, including some wattles, Swamp Paperbark (*Melaleuca ericifolia*) and Radiata Pine (*Pinus radiata*).

Fresh Koala scats smell of eucalyptus oil, but when dry they have little odour. They are found on the ground beneath the food trees, and beneath the forks the Koalas sleep in.

Key to Scats No 15, Plate 30, page 103

Wombats

Scats of all three wombat species are remarkably similar, not only in shape and size but also in colour and texture. They are usually produced in groups of four to eight separate, cube-shaped pellets.

The surface is dark and shiny with a mucus covering when the scats are fresh, and this dries to a dark brown or black casing. The scats are green or green-brown when broken, and the plant material is uniform and close-textured. They resemble the scats of the Eastern Grey Kangaroo in this respect, and oddly shaped Common Wombat scats can easily be confused with scats of that species.

Fresh wombat scats smell strongly of wombat — an unmistakable sweetish, peaty odour which, like all smells, is difficult to describe. It lingers in old scats but is less strong.

Common Wombat scats are often deposited in conspicuous places — on rocks, fallen branches or heaps of earth — and are also scattered about where the wombat has been feeding (Figure 219 and Figure 211, page 85).

Scats of the two species of hairy-nosed wombats are found near their burrows and along the paths leading to them, as well as at their feeding areas (Plate 84, page 215). Large numbers of scats accumulate at rubbing posts and dust wallows (Plate 86, page 215).

Key to Scats Nos 83–85, Plate 43, page 130

165

BANDICOOTS AND BILBY

Scats are produced as firm, cylindrical pellets with a dark, shiny coating that dries to a smooth, mud-like surface. They are very friable and are often found broken into two or more pieces.

Bandicoots and Bilbies eat mainly insects and other invertebrates, and fine particles of these can be seen in a broken scat. Soil is ingested as the invertebrates are dug up from the ground, and the soil particles colour the scats brown or black. Sand in the pellets of the Bilby often colours them red-brown. The Bilby and most bandicoots eat some plant material as well as insects, so fibrous material is also found in their scats.

Fresh scats have a meaty smell, but this fades with age, and dry scats have little odour. Bandicoots and the Bilby often deposit their scats near the conical holes they dig while searching for food (see Plate 90, page 216).

Key to Scats Nos 11–14, 21 Plates 29, 31, pages 102, 106

MARSUPIAL CARNIVORES

Quolls, Tasmanian Devil and some smaller relatives
Cylindrical scats, usually twisted and often pointed at one end, are typical of the larger marsupial carnivores, but there is much variety in their shape and size.

As they are carnivorous, these species' scats contain such items as fur, bone fragments, feathers and reptile scales. They also eat insects, and most of them are opportunists, raiding litter bins for garbage when this is available.

Tasmanian Devil scats vary from tightly twisted cylinders, generally composed of fine fur, to drier scats which may be found broken into several pieces. They are usually grey because of the digested bone, and large, sharp bone fragments are often present.

Spot-tailed Quoll scats are often dark brown with an oily texture, but they vary with the diet. They are twisted when composed of fur, but may be straight cylinders when the quoll has been eating non-mammalian species or plant matter. Scats of the Eastern Quoll are not as oily or pungent as those of the Spot-tailed Quoll. However, the size range overlaps with that of small Spot-tailed Quolls, and the structure also varies from large twisted scats full of fur to straight cylinders packed with beetle casings, and it is difficult to distinguish between the scats of these quolls.

All the scats of these animals have strong, unpleasant odours, unique to each species, but as with all smells they have to be personally experienced in order to tell the difference. The odour of marsupial carnivore scats is generally different from the odour of the introduced carnivores (Dog, Red Fox and Cat) and is a useful way of distinguishing between them.

The Spot-tailed Quoll and the Western Quoll often leave scats repeatedly at the same latrine site. These are usually conspicuous places such as rocks or boulders. The Northern Quoll's latrines are also often in prominent places, such as rock piles or boulders, usually on the highest point available, such as ridge-tops or hills. This quoll has been found to occasionally deposit scats at the den entrance (see Plate 94, page 217). Meri Oakwood, studying these quolls in the Northern Territory, has observed that scats seem to be deposited at den entrances only during October and November, at a time when the young are beginning to forage for themselves. It is likely that the scats act as recognition signals for the young quolls. Where quolls live near human habitation, they often leave scats conspicuously on house verandahs, bare garden plots, driveways and so on. Scats of some snakes can be mistaken for quoll scats (see Figures 232, 233).

Scats of the Tasmanian Devil are also found at latrine sites — commonly at track junctions and creek crossings. They are also deposited along tracks, and there is sometimes a concentration of scats near a carcass.

The smaller carnivores, such as the Kowari and the Mulgara, leave their strong-smelling, twisted scats near their burrows and on rocks as territorial markers. The scats of some snakes can be mistaken for those of these small carnivores (see Figures 232, 233).

Key to Scats Nos 5, 6, 8–10, 31, 33, Plates 27, 28, 33, pages 98, 99, 110

Phascogales and Antechinuses
The cylindrical scats of the phascogales and the larger antechinuses are narrow twists of fur or feathers, or are sometimes made up entirely of insect remains. The scats of the smaller species, consisting entirely of insect remains, are very friable. All these scats have a strong odour when fresh, and in the larger species the smell lingers in dry scats.

Scats of the Brush-tailed Phascogale may be found near their dens, on tree branches and fallen logs. Scats of the smaller phascogales and those of most of the antechinuses are difficult to find in leaf litter, but some species live near human habitation and their scats may be found in sheds and other buildings.

Key to Scats Nos 30, 32, 50, 51, 52, Plates 33, 36, 37, pages 110, 115, 118

Dunnart, Planigales and Other Small Dasyurids

The small, twisted cylindrical scats of these animals are similar to those of the smaller antechinuses. They contain fine fragments of insects and are very friable. They have a strong odour when fresh, but little or no smell when dry.

Scats of these small species are rarly found, but thtey are deposited in large numbers near their nest sites, which are also difficult to find. The White-footed Dunnart has been found nesting under sheets of bark on the ground, with scats in small piles also under the bark covering. The Common Dunnart also nests on the ground under sheets of bark, and some man-made objects, such as sheets of roofing iron.

Key to Scats Nos 53–57, 68, 69, Plates 37, 39, pages 118, 122

Numbats

The scats of the Numbat are smooth, dark cylinders containing fine fragments of termites and soil particles. They are coated with mucus that dries to a firm glaze. The meaty odour of the fresh scats is similar to that of bandicoot scats, but older scats have little odour.

Numbat scats are often found on fallen logs, and sometimes on the ground where the animal has been digging shallow holes in its search for termites in the galleries radiating from a termite mound (see Plate 98, page 219).

Key to Scats No 22, Plate 31, page 106

Marsupial Mole

Scats of this species were not available for examination.

RODENT FAMILY

Rats and Mice

Most rodents produce narrow, cylindrical scats, usually with one or both ends pointed. They vary in size from the relatively large Giant White-tailed Rat's pellets to the tiny pellets produced by the House Mouse and other small mice.

As the diets of the many rats and mice vary from species to species and place to place, the scat contents vary also. Fine, powdery particles of plant material, pollen grains, and insect fragments are found in many rodent scats. Traces of feathers or fur, fragments of egg shells, and reptile scales are other items occasionally found in rat scats.

The odour of some rodent scats can be a useful diagnostic aid. Bush Rat scats, for example, have a characteristic fetid odour, while the Black Rat's scats have a different, but also characteristic, smell. House Mouse scats have the same strong, musty smell as the mice themselves.

The scats of most rodents are dark brown or black, but a few species produce lighter-coloured material. The Broad-toothed Rat's scats, which are greenish-yellow when fresh but yellow-brown when dry, can be used to confirm the presence of this species in an area. They are easily distinguished from the darker scats of Bush Rats, which often use the same runways as the Broad-toothed Rat. The pale brown of the Desert Mouse's scats distinguishes them from the darker scats of other desert rodents, such as the Sandy Inland Mouse.

Scats of rats and mice are found along their runways and at the entrances to their nests or burrows. Scats of rats and mice living in or near human habitation are found in dark corners of rooms and sheds,and other similar places.

Key to Scats Nos 34–45, 58–61, 64–67, 73–75, Plates 34, 35, 38, 39, 40, 41, pages 111, 114, 119, 122, 123, 126

Water-rat

Water-rat scats are cylindrical pellets, often broken into one or more pieces. They contain fairly coarse fragments of various food remains. Insect casings, fish scales, shell fragments, feathers, hair and bones have all been found in the scats. They have a strong, unpleasant odour. False Water-rat scats were not available for examination.

Scats of the Water-rat are found at or near the feeding sites, or 'dining-tables'. These are often flat rocks or stumps near water, on jetties, piers or moored boats, or on the ground in small sheltered clearings near the shores of rivers and lakes (Figure 220; see also Plate 106, page 230)

Key to Scats No 29, Plate 33, page 110

Figure 220
A Water-rat's
feeding table

Hopping-mice

Very small cylindrical pellets are produced by all species of hopping-mice available for examination. They contain very fine particles of plant material, and have a musty odour when fresh. Scats of the hopping-mice are found scattered near their burrows and runways.

Key to Scats Nos 70–72, Plate 40, page 123

RABBIT AND HARE

Rabbit scats are round, slightly flattened and usually dark and moist when fresh, drying out to dark or light brown, depending on the vegetation consumed. They contain fine particles of plant material and are usually less than 1 cm wide. Hare scats are also round and flattened, but are generally larger than rabbit scats and contain coarser plant material. Both rabbit and hare scats smell of herbs or grasses.

Rabbits generally deposit their scats on 'hills', which may be slightly elevated patches of ground, ant hills or rocks, where many scats will accumulate. These hills are used as territorial markers (Figure 221).

Rabbit scats are also found on the runs near the burrows, and a few scats are often deposited on the pile of loose earth at the end of the small 'scrapes' dug in the ground (Figure 238, page 228). A hare does not use regular latrine sites but leaves scats where it has been feeding and near its shelter.

Key to Scats Nos 127, 128, Plate 55, page 154

*Figure 221
Many rabbit
scats may
accumulate on a
'hill'*

171

INTRODUCED CARNIVORES

Dingo, Dog, Red Fox and Cat

The cylindrical scats of a fox, a small Dog, and a Cat are similar in many ways, as are those of a large Dog and a Dingo. In general, a fox's scat is no wider than 2 cm and a Cat's is between 2 and 2.5 cm in diameter. Dog scats vary in size with the breed of the Dog. The scats of Dingoes and large feral Dogs may be as much as 4 cm in diameter.

There is much variation in the colour and shape of predator scats, just as there is in the diet of these animals. Long, slightly twisted cylinders, often with a tuft of hair or whiskers visible at the pointed end, are the most characteristic scats of Dogs and Red Foxes. Sometimes they are sausage-shaped, and may be grey or white as a result of the bone material eaten (Figure 222). Fur and bone fragments, feathers, reptile remains, fish scales and bones, and yabbie shell fragments are all commonly found items in predator scats (see Figure 214, page 89).

Although generally carnivorous, Dogs and Red Foxes will eat plant material such as berries and fruits, and insects, such as beetles and large moths, when these are plentiful. Their scats are sometimes composed almost entirely of blackberry seeds or moth wings. Cats also eat insects and plants at times. All three predators raid garbage bins and rubbish tips when these are available, and it is not uncommon for pieces of plastic bags, silver paper and similar items to appear in their scats.

The main difference between the three kinds of scats — Dog/Dingo, fox and Cat — is their odour, each one having its own characteristic smell, all of them strong and unpleasant. Only personal experience can enable you to distinguish between these odours. Also, Dog scats tend to have larger fragments of bone material in them than fox scats. Sometimes these are visible on the surface of the scat.

Dogs, Dingoes and foxes often deposit their scats on elevated places such as rocks or grass tussocks. They may use the same site more than once; intersections and corners of tracks and roads are frequently used as defecation sites. Sometimes one Dog or fox deposits its scats on top of those of another. Scats are also commonly found near the bodies of dead animals: road kills, cattle carcasses and so on.

Cats usually scratch sand, soil or leaf litter over their scats to cover them, and they often use the same sites regularly. In sandy areas, large numbers of Cat scats may be found at a site where the sand has blown away.

Key to Scats Nos 1, 2, 7, Plates 25, 27, pages 94, 98

Figure 222   Some Dog and Red Fox scats are very similar in shape and size

HOOFED MAMMALS

Horse and Donkey

The scats of the Horse and the Donkey are similar: large brown, rounded pellets, often still partly joined to one another, deposited in heaps. Undigested hay or grass stems can often be seen in the scats. Fresh scats have a strong odour, but this fades in drier material.

Scats are deposited where the animals are feeding (Figure 223), and stallions and male Donkeys will regularly leave scats in the same places until large piles have accumulated, presumably as territorial markers.

Figure 223 (Left) Donkey scats

Figure 224 (Right) Clumps of sheep scats

Deer, Sheep and Goat

Deer, sheep and Goats produce rounded, oval or cylindrical pellets. These may be deposited as single pellets but are more often produced in clumps, which may fall apart as they hit the ground and scatter as individual pellets, or remain as clumps. The shapes not only vary from species to species, but sometimes also within a species. Some deer, for example, may produce differently shaped pellets at different seasons. Sheep and Goat scats, which are often found in the same areas, have many similarities. A team from the CSIRO has devised simple criteria for distinguishing between some of the scats produced by sheep and Goats in mulga scrub in north-western New South Wales. Figure 225 shows some of their criteria. In wetter country, individual sheep pellets have angled facets, probably because they are more distorted in the wetter clumps (Figure 224).

All the scats of deer, sheep and Goats contain fairly fine plant material, and they have a strong, grassy odour when fresh. They are found where the animals have been feeding and moving.

Key to Scats Nos 120–126, Plates 54, 55, pages 151, 154

Identify samples as 'goat' if both ends of most pellets are pointed or *if the pellets are cylindrical*

Identify samples as 'sheep' if either end of most pellets is dimpled or if both ends are round

Identify samples as 'unknown' if most pellets have one pointed end and one round end and if pellets are spherical to cigar-shaped

Figure 225   Identification of sheep and goat scats, using criteria devised by the team from the CSIRO.

Figure 226
Cattle pats

Cattle and Buffalo

Scats of these animals are deposited either as a brown, semi-liquid mass which dries out in the form of one large flat 'cake' or 'pat', or (when their food is drier) several smaller, firmer, layered pats (Figure 226). Fresh scats have a strong, characteristic odour that fades when the scats dry out.

Scats are deposited where the animals are feeding. If an animal is on the move, a trail of small pats may be found.

Camel

Large, rounded pellets, often joined in pairs, are produced in large numbers by camels. They contain coarse plant material. Fresh scats have a strong, acrid odour. They are dropped where the animals are feeding, and along their trails.

Key to Scats No 82, Plate 42, page 127

Pig

When they are eating only plant material, Pigs produce roughly cylindrical brown scats made up of smaller, round, flat 'cakes' that are joined together to form the cylinder. Feral Pigs have a varied diet, and their scats may contain plant material, fur and bones, feathers and so on.

Depending on the diet, feral Pig scats have either a strong peaty smell or a most unpleasant acrid odour.

Scats are found where the Pig has been feeding. Areas of disturbed ground where a pig has been rooting in the ground for larvae will often have pig scats nearby (Plate 112, page 232).

Key to Scats No 81, Plate 42, page 127

MARINE MAMMALS

Fur-seals and sea-lions produce similar scats. These are large and roughly cylindrical, and set like concrete when they dry. These animals also produce wet, shapeless waste material when they have been feeding on soft-bodied animals.

Scats of these animals have a very strong, unpleasant, fishy odour when fresh. They are found on the rocks and sandy places where these animals haul out.

Scats of the Leopard Seal are occasionally found on ocean beaches. These scats are large and usually contain the feathers of sea birds.

Key to Scats No 4, Plate 26, page 95

BATS

Large bats, such as the flying-foxes, produce scats that accumulate on the ground at their camps, forming a shapeless mass. Firm scats are occasionally produced, but generally the soft fruit in the diet of these bats makes the waste material wet and formless.

The scats of each species of flying-fox are said to have a strong, individual odour. They are found on the ground at flying-fox camps and where the bats have been feeding.

The scats of microbats are very small and friable. In most species they are composed entirely of wings and other fragments of insects. They have a sweetish smell when fresh, and are found where these bats roost. For example, they collect in large numbers on the floors of caves and old mine shafts. They are also found in buildings where small bats live in or near human habitation.

Key to Scats Nos 79, 80, Plate 41, page 126

Other Animals

Although it is usually easy to tell the difference between the scats of mammals and those of birds or reptiles, there are some that are similar. Also, some insects, particularly in the larval stage, produce waste material in the form of pellets that could be mistaken for the scats of small mammals.

Birds

Some bird species regurgitate much of the indigestible parts of their food in the form of pellets or castings. These pellets are similar to some carnivore scats, because they may contain items such as bone and fur, feathers, shells, insect castings and plant remains, particularly seeds. The major difference between bird pellets and mammal scats is that there is usually a considerable amount of powdery material, as well as fur and bones and so on, in a scat. Pellets have very little or no powdery residue.

The size and shape of bird pellets vary widely. Birds that commonly produce pellets include eagles, hawks, owls, gulls, some waders, magpies, ravens and currawongs. The Pied Currawong has a wide-ranging diet, and a variety of its pellets are often found near human habitation. Some bird pellets are shown in Plate 56, page 155 and Figures 229, 230, 231).

Most bird scats are easily distinguishable from those of mammals, as they have a cap or smear of white material attached to them. Birds concentrate their urine and excrete it as a white substance, uric acid. This substance is very friable and often breaks off the scat or disintegrates to a powder, so the remaining cylindrical scat, such as those of swans, ducks or geese, could be mistaken for mammal scats (Figure 227). The scats of perching birds often pile up in small coils, while those of emus form conical piles when these birds have been eating dry vegetation (Figure 228). When the vegetation is wet, emu scats are formless splashes of greenish paste.

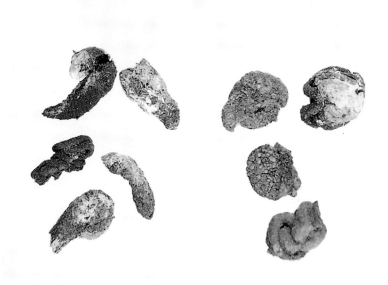

Figure 227   Assorted bird scats

Figure 228   Assorted Emu scats

179

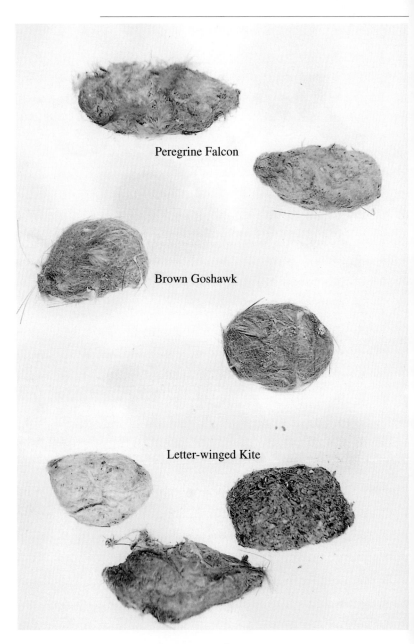

Peregrine Falcon

Brown Goshawk

Letter-winged Kite

Figure 229   Regurgitated pellets of some birds of prey

White-breasted Sea-eagle

Wedge-tailed Eagle

L. ADDISON

Figure 230   Regurgitated pellets of large eagles

181

Australian Magpie

Rainbow Bee-eater

Pied Currawong

Figure 231   Regurgitated pellets of birds

Reptiles

Many reptile scats, like those of birds, have a capping of white material at one end, and this also often becomes detached, leaving scats very similar in shape and size to those of certain mammals (Figures 232, 233).

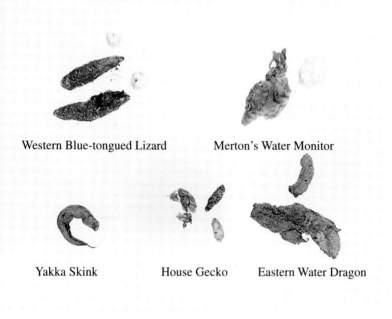

Western Blue-tongued Lizard   Merton's Water Monitor

Yakka Skink   House Gecko   Eastern Water Dragon

Figure 232   Assorted reptile scats

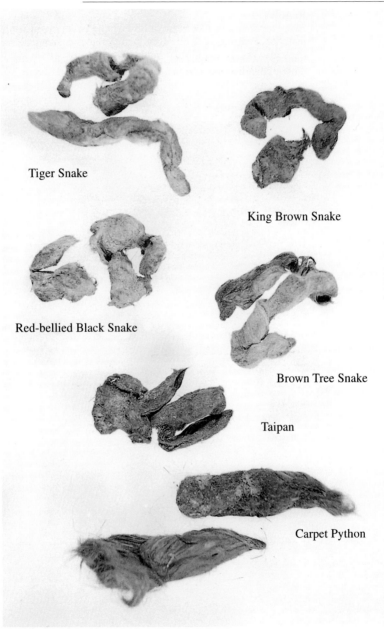

Tiger Snake

King Brown Snake

Red-bellied Black Snake

Brown Tree Snake

Taipan

Carpet Python

Figure 233   Assorted reptile scats

Amphibians

Scats of the smaller frogs and toads are not often found, but Cane Toad scats are large and often left in prominent positions, such as rocks and boulders. Some Cane Toad scats resemble those of bandicoots, having a similar shape, size and content of insect material, but toad scats generally have a rougher surface (Figure 234)

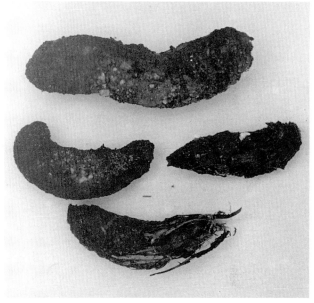

Figure 234   Scats of the Cane Toad

Invertebrates

Most of the larvae of large moths and butterflies are herbivorous, and their waste material contains fine particles of plant material. Some are wood-borers, as are some beetle larvae. Their waste material has some similarities to the scats of some small mammals, such as rodents. Frass is often found in rotting wood and in tree-holes. Many other insect larvae, and sometimes the adults as well, produce waste that can be mistaken for mammal scats (Figure 235).

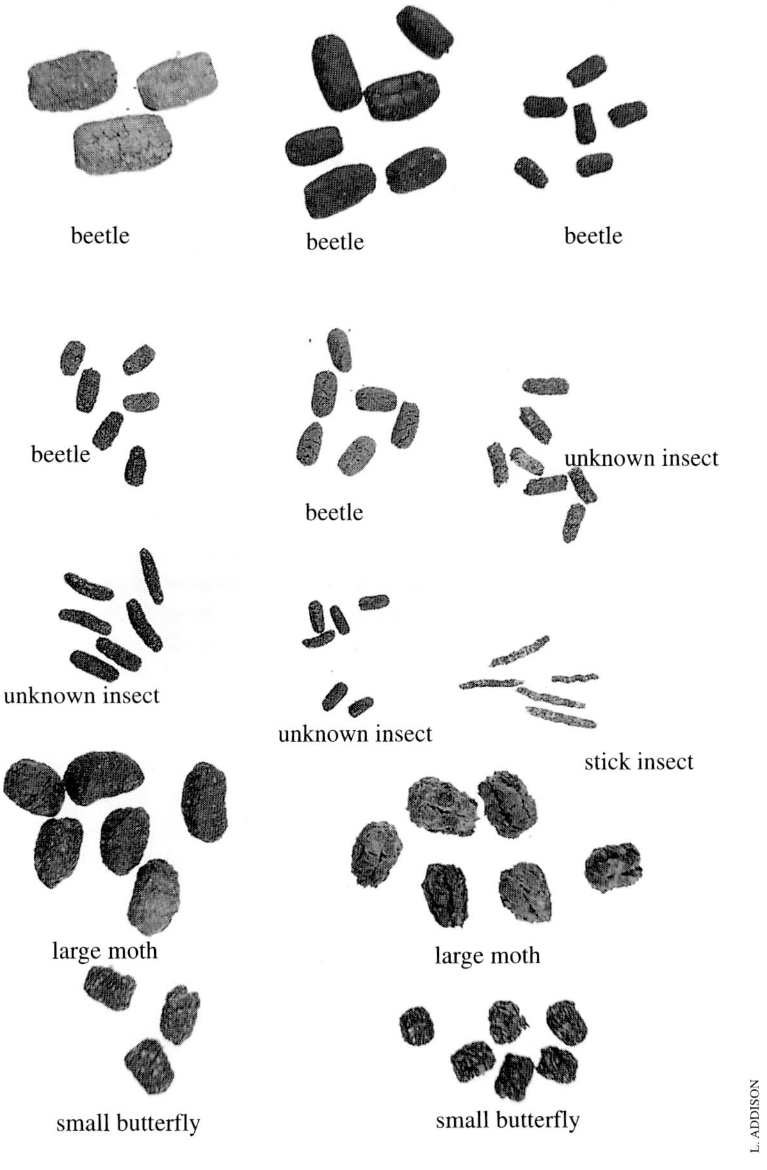

beetle

beetle

beetle

beetle

beetle

unknown insect

unknown insect

unknown insect

stick insect

large moth

large moth

small butterfly

small butterfly

L. ADDISON

Figure 235   Assorted scats of insect larvae

SHELTERS, FEEDING SIGNS AND OTHER TRACES

Tracks and scats are not the only signs that show us that mammals have been in an area. The landscape abounds with evidence of their activity in the form of dwelling places, feeding signs and many other traces.

Although many mammals do not have permanent homes that they use throughout the year, some construct several shelters and use them frequently. Burrows and nests of many shapes and sizes can be found, and their owners can be identified.

Many animals do not build shelters, but make use of natural features such as hollow trees and caves. Some mammals use a variety of sheltering places, and some of these are used by more than one species, so it is often difficult to be certain of the owner of a shelter without additional evidence, such as scats and tracks.

Many mammals leave obvious traces in areas where they have been feeding. Often these will be found while you are following a trail: a fox's track, for instance, may lead to the site of a kill or a cache of food; a wombat's trail may show where it has stopped to graze on roadside tussocks.

Other feeding signs can be found on the ground. Holes, large and small, are dug by many species as they search for food such as underground fungi or the larvae of beetles and other insects. Echidnas, bandicoots, potoroos, Numbats and wild Pigs are just some of the species that leave signs of their digging.

Sometimes remains of the food itself may be found: a scatter of blossoms may show that a flying-fox was in the canopy above; empty or broken seed cases, gnawed fruit and broken egg shells are a few other examples of feeding signs.

The trunks and branches of trees and shrubs also often show marks where a mammal has been feeding. Slashes made in the bark, from which sap oozes, are used as feeding sites by several species of gliders, and these slashes can sometimes be identified years after they were made. Some possums also gouge wood-boring grubs out of the trunks of wattles and other trees.

Tree trunks also show other marks of an animal's passage: scratch marks left by the sharp claws of possums and koalas, regular landing places of gliders, and rubbing posts and trees 'marked' by dogs and foxes are some examples.

There are other traces on the ground that can be used to identify the species responsible. Large areas of disturbed ground may be the sites of wallows, and characteristic scrapes and scratches are made by several species.

Various animals use definite pathways or runs that can often be found; the animal can be identified by the kind of run it has made, or by the other signs it has left on the pathway.

Key to Shelters

As many mammals use a variety of shelters, and some shelters are used by more than one species, the following key is not comprehensive and is intended as a general guide only. It refers to the traces mentioned in the text. Many other traces of mammals can be found. The numbers in parentheses are the page numbers to which you should refer for each animal.

A BURROWS
1 **With entrance more than 20 cm in diameter:**
   wombats (211), Tasmanian Devil (222), Dingo and Dog (237), Red Fox (237)
2 **With entrance between 10 and 20 cm in diameter:**
   Platypus (194), Burrowing Bettong (205), Bilby (221), quolls (222) Water-rat (226), European Rabbit (228)
3 **With entrance less than 10 cm in diameter:**
   Kowari (222), Mulgara (222), antechinuses (223), Numbat (224), Marsupial Mole (224), many rats and mice (225)

B UNCONSTRUCTED RESTING PLACE
1 **In open grassland or woodland:**
   kangaroos (195), some wallabies (195), Brown Hare (228), some hoofed mammals (239)
2 **In thick scrub or dense undergrowth, or under logs:**
   Short-beaked Echidna (193), some wallabies (195), pademelons (196), Quokka (196), quolls (222), Dingo and Dog (237), Pig (239), deer (239)
3 **In grass tussocks, spinifex hummocks, or among rocks:**
   hare-wallabies (196), potoroos (205), Kultarr (223), planigales (223), ningauis (223), Brown Hare (228)
4 **In tree forks and branches:**
   Koala (211), tree-kangaroos (196), Spotted Cuscus (206), Green Ringtail Possum (207), flying-foxes (241)
5 **In clumps of mistletoe, epiphytic ferns, or dense bushes:**
   Mountain Brushtail Possum (206), Common Ringtail Possum (207), Herbert River Ringtail Possum (207), Daintree River Ringtail Possum (207), Lemuroid Ringtail Possum (207), some megabats (207)

C   IN ENCLOSED SHELTERS WITH NO CONSTRUCTED
NEST

1   **In caves, rock fissures or old mines:**
Short-beaked Echidna (193), wallaroos (195), rock-wallabies (196), Scaly-tailed Possum (206), Rock Ringtail (207),
quolls (222), Tasmanian Devil (222), Feral Goat (239), Bare-
backed Fruit-bat (241), Ghost Bat (241), many microbats (241)

2   **In hollow logs and stumps, on or near the ground;**
Short-beaked Echidna (193), quolls (222), Tasmanian Devil
(223), Kultarr (223), Dingo and Dog (237), Red Fox (237),
Cat (237)

3   **In hollows in tree-trunks or limbs:**
brushtail possums (206), Grey Cuscus (206), quolls (222),
Red Fox (237), Cat (237)

4   **In or under houses, sheds and other buildings:**
brushtail possums (206), quolls (222), Tasmanian Devil (222)

5   **In termite mounds:**
Northern Brushtail Possum (206), Northern Quoll (222), Cat
(237)

D   IN ENCLOSED SHELTERS WITH A CONSTRUCTED
NEST

1   **In tussocks or undergrowth or among rocks:**
bettongs (205), Musky Rat-kangaroo (205), Little Pygmy-
possum (210), bandicoots (221), some antechinuses,
dunnarts, planigales and ningauis (223), many rats and mice
(225)

2   **In hollow logs or stumps:**
Little Pygmy-possum (210), Long-tailed Pygmy-possum
(210), bandicoots (221), phascogales (223), some
antechinuses and dunnarts (223), some rats and mice (225)

3   **In hollows in tree-trunks or limbs:**
Common Ringtail Possum (207), gliders (208), Striped Possum (208), Leadbeater's Possum (208), Eastern Pygmy-possum (210), Western Pygmy-possum (210), Feathertail Glider
(210), phascogales (223), some antechinuses and dunnarts
(223), some rats and mice (225), tree-rats (226), Brush-tailed
Rabbit-rat (226)

4   **In bulky nest in tree-forks and branches:**
Common Ringtail Possum (207), Herbert River Ringtail Possum (207)

5   **In abandoned birds' nests:**
Feathertail Glider (210), Brush-tailed Phascogale (223)

6   **In the walls and roofs of houses, sheds and other buildings:**
Common Brushtail Possum (206), Northern Brushtail Possum
(206), Common Ringtail Possum (207), some antechinuses
(223), some rats and mice (225)

191

Key to Feeding Signs and Other Traces

A AREAS OF DISTURBED GROUND
1 **Wallows and water-digs:**
   kangaroos (195), wallaroos (195), wombats (212), Water Buffalo (239), feral Pig (239), deer (239)
2 **Scratchings and diggings:**
   echidna (193), bettongs (205), potoroos (205), bandicoots (221), Bilby (221), Numbat (224), Common Wombat (212), European Rabbit (228), Dingo and Dog (237), Red Fox (237), Cat (238)

B FEEDING SIGNS
1 **On tree-trunks and branches:**
   wallabies (195), gliders (208), Striped Possum (208), European Rabbit (228), Brown Hare (228)
2 **On leaves and flowers:**
   ringtail possums (207), gliders (208), flying-foxes (241)
3 **On fruits and seeds:**
   brushtail possums (206), ringtail possums (207), Mountain Pygmy-possum (210), rats and mice (226)
4 **On prey, or prey remains:**
   Tasmanian Devil (222), phascogales (223), antechinuses (223), dunnarts (223), Water-rat (227), Dingo and Dog (238), Red Fox (238), Cat (238), Ghost Bat (241), microbats (241)

Guide to Shelters, Feeding Signs and Other Traces

SHORT-BEAKED ECHIDNA

Shelter

Echidnas shelter in hollow logs, rabbit or wombat burrows, rock caves, thick vegetation or piles of forest litter. When caught in an exposed position on soft soil or sand they will sometimes rapidly dig straight down, apparently sinking into the ground, leaving just a few spines visible.

In the breeding season, which is usually in winter, the female may dig a burrow or refurbish an old one in which to put her young. These burrows have been found at the bases of termite mounds, in the soil between the roots of trees, and occasionally in mounds of dense vegetation. They are about a metre long, with an enlarged nursery chamber either at the end or in a short side tunnel. These burrows are hard to find, as the entrance is plugged and well hidden. Some females use a burrow while incubating the egg, but others simply shelter in secluded places above ground.

Feeding signs

The nests of ants and termites often show the damage caused by a feeding echidna. These nests, which may be under ground level, are broached with a conical hole up to 20 cm deep, with the mark of the snout at its end (Plate 57, page 197).

Ant and termite mounds above the ground often show larger, echidna-sized holes, again with snout marks, (Plate 58, page 197).

Sometimes tunnels up to a metre deep are found, penetrating the mounds. These tunnels have a rounded roof and a flat floor. Although echidnas usually only burrow into mounds in late winter and spring, the damage is obvious for a long time.

Rotting wood on the ground which has been disturbed and torn apart is often the work of an echidna's strong forepaws as it searched for food, which includes soft material such as invertebrate larvae and eggs, worms and so on. Small scratches on the ground may be found where leaf litter has been turned over, and sometimes larger areas of soft soil or pasture are turned over by the snout in a corkscrew action, leaving damage similar to that made by bandicoots.

Other signs

During courtship, a female echidna may be followed by a 'train' of several males. Dr Peggy Rismiller, of the Pelican Lagoon Research Centre on Kangaroo Island, has found that males attempting to breed with the female will compete with one another by digging trenches next to the female so that they can lift her up, and place their tails under hers in attempts to mate with her. One male pushes all the others out before he succeeds in mating. The trenches dug by the males leave a distinctive ring in the dirt (Plate 59, page 197).

PLATYPUS

Shelter

The Platypus digs a burrow, usually about 6 or 7 metres long, with the entrance on a sloping bank, above the water-line at heights varying from a few centimetres to several metres. The entrance has a low, arched shape; it is about 10–12 cm wide and 8 cm high, providing a tight fit for the platypus as it enters the burrow (Plate 60, page 197).

The entrance is often hidden behind reeds or other vegetation. Soil excavated from the burrow is packed flat against its sides, and any that spills out is flattened by the body and tail of the Platypus. Whether or not a burrow is in use can often be decided by the state of the 'slide' of flattened mud outside the entrance.

Feeding signs

As it feeds in the water, the Platypus leaves no particular feeding signs.

Other traces

The 'slide' of wet, flattened mud outside a burrow entrance, and sometimes on other parts of creek and river banks, are indications that a Platypus regularly uses those places to enter and leave the water.

KANGAROOS, WALLAROOS AND
LARGE WALLABIES

Shelters

The kangaroos and other large macropods do not have permanent shelters, and their resting places are not very well defined.

Kangaroos will often rest in the shade, usually under trees and shrubs, often choosing a place where their backs are protected but they have an open view of the other three sides. In sandy country they may scrape out a resting place in the sand. Wallaroos often rest in caves and on rocky ledges.

Wallabies may regularly use the same resting place. Patches of well-flattened grass in protected situations, such as among ferns or in thick scrub, often indicate where they have been sleeping.

Feeding signs

By following a macropod's trail it is sometimes possible to see where it has been grazing on tussocks. Some wallabies browse on shrubs and the lower branches of some trees. This can result in the stripping of the leaves and smaller twigs from the lower branches, leaving them dead (Plate 61, page 198).

Other traces

In the dry, desert country of central Australia, kangaroos and wallaroos will scoop out the sand on dry watercourses to find water, leaving characteristic 'water-digs' (Plate 62, page 198).

Male kangaroos will sometimes scratch deep gouges in tree trunks with their powerful front claws, as part of their aggressive displays to other males. They will also grasp small shrubs and tussocks, sometimes pulling them out by the roots, and these damaged plants can sometimes be found where the kangaroos have been active.

Well-worn pathways in the forest often lead to the sleeping places of wallabies.

PADEMELONS, ROCK-WALLABIES, TREE-KANGAROOS, NAILTAIL WALLABIES, HARE-WALLABIES, QUOKKA

Shelters

Pademelons rest in thickets and other dense vegetation. They regularly use the same resting places, producing small hollows in the vegetation.

Rock-wallabies shelter in caves, on rock ledges, in fissures and crevices in the rocks, and sometimes in dense vegetation such as Lantana. Nailtail wallabies scrape out a shallow depression in the shade, usually beside a bush or tree. Tree-kangaroos rest, sitting crouched, in the crown of a tree or on a branch.

Hare-wallabies shelter in hides or 'squats' under large spinifex hummocks, grass tussocks or shrubs. The squats are about 30–50 cm long and about 20–25 cm high. Quokkas shelter in groups in dense vegetation.

Feeding signs

Apart from cropped grass and some evidence of browsing on low bushes, the small macropods leave no particular feeding signs.

Other traces

Well-worn runways in dense vegetation may lead to the resting places of pademelons and quokkas.

Patches of smooth rock may give a clue to the presence of rock-wallaby shelters, as the rocks are often worn smooth by the constant passage of the rock-wallabies' feet.

Deep scratches on the trunks and branches of trees may be found where tree-kangaroos are present, made by the strong, recurved claws and forelimbs as these grip the tree in climbing. Dried patches of mud on the trunks may be the remnants of muddy footprints of these animals.

RISMAC

RISMAC

Plate 57   Shallow echidna dig, with a nose-poke at the end (page 193).

Plate 58   Echidna digs in a meat ant mound (page 193).

RISMAC

FISK

Plate 59   Dr Rismiller examining the mating trench of echidna (page 194).

Plate 60   Entrance to a Platypus burrow on a creek bank (page 194).

197

Plate 61   Dead lower branches are evidence of wallaby browsing (page 19⁵.

Plate 62   Kangaroo water-dig (page 195).

K. ATKINSON

A. WOOLNOUGH

Plate 63   Small conical pit dug by a Northern Bettong on firm ground (page 205).

J. FRIEND

Plate 64   A Brush-tailed Bettong dig (page 205).

Plate 65   Conical pit dug by a Long-nosed Potoroo, with a remnant of the fruiting body of an underground fungus nearby (page 205).

Plate 66   Sporocarps of underground fungi are eaten by bettongs and potoroos (page 205).

A. CLARIDGE

R. INGAMELLS

Plates 67   Tree hollow used as a den site by the Common Brushtail Possum (page 206).

Plate 68   A lemon shows the mark of the teeth of a brushtail possum (page 206).

Plate 69   The soft bark of this tree shows the marks of many possum visits (page 206).

Plate 70   The bulky nest (drey) of a Common Ringtail Possum (page 207).

Plate 71   A scatter of wattle foliage on the ground is evidence of Common Ringtail activity (page 207).

Plate 72   This old stag has numerous fissures and hollow limbs and provides dens for several families of Sugar Gliders (page 208).

Plate 73   Vertical gashes made by Sugar Gliders in the trunk of a Rough-barked Angophora (Angophora floribunda) (page 208).

Plate 74   Incisions made by Sugar Gliders on the trunk of a Yellow Gum (Eucalyptus leucoxylon) (page 208).

Plate 75   Hole used by Leadbeater's Possum in an old stag of Mountain Ash (Eucalyptus regnans). The edges of the hole are smoothed and rounded by the possums (page 208).

Plate 76   Gum of Silver Wattle (Acacia dealbata), which is eaten by Leadbeater's Possum (page 208).

Messmate *(E. obliqua)*

Mountain Ash *(E. regnans)*

Southern Blue Gum *(E. globulus)*

Mountain Grey Gum *(E. cypellocarpa)*

Silver Top Ash *(E. sieberi)*

Apple Box *(E. bridgesiana)*

Red Bloodwood *(E. gummifera)*

Mountain Gum *(E. dalrympleana)*

Plate 77   Sap feeding sites of Yellow-bellied Gliders (page 208).

Plate 78   Gashes made by Yellow-bellied Gliders many years ago leave characteristic scars (page 208).

203

Plate 79 'Chew balls' of bark made by Yellow-bellied Gliders at feeding sites (page 208).

Plate 80 An Eastern Pygmy-possum nested in this stump of a dead grass-tree (Xanthorrhoea sp.) (page 210).

BETTONGS, POTOROOS AND
RAT-KANGAROO

Shelters

All species of bettongs build shelters. The rare Burrowing Bettong digs complex burrows, often with many entrances, but all other species build nests. These are made of plant material such as grass and shredded bark, and are oval or cone-shaped, with an entrance at one end about 5–10 cm in diameter. The nest is constructed over a shallow excavation, often at the base of a tussock, under a shrub or a fallen log, or amongst dense undergrowth. Bettongs usually have several nests that they use regularly.

Potoroos probably do not construct elaborate nests. They scrape out a depression under a shrub or tussock, and this may be roughly lined with plant litter.

The Musky Rat-kangaroo makes a nest of dried leaves and ferns, usually in a clump of Lawyer Vine.

Feeding signs

Bettongs and potoroos all make small, roughly cone-shaped holes in the ground as they dig for fungi and invertebrate larvae. The size and shape of these diggings depends on the nature of the ground: in firm soil they are generally conical pits with rounded bottoms. They vary considerably in size, but may be as much as 25 cm deep or as little as 6 cm (Plates 63, 64, 65, page 199).

On softer ground the holes are often larger and less well defined. Sometimes a number of holes will be found together: the Rufous Bettong often digs up large areas of ground in its search for beetle larvae.

At certain seasons, when the fruiting bodies (sporocarps) of underground fungi are the main source of food, bettongs and potoroos will occasionally leave the remains of these fruiting bodies near their diggings (Plates 65, 66, page 199).

Bandicoot diggings, often found in the same areas as those of bettongs and potoroos, tend to be smaller, narrower and generally have a more pointed end, but it is often very difficult, if not impossible, to distinguish between them (see page 221, and Plate 90 on page 216.)

BRUSHTAIL POSSUMS, SCALY-TAILED POSSUM AND CUSCUSES

Shelters
All the brushtail possums prefer to use hollow limbs in trees as their dens (Plate 67, page 200), but when there are no hollows available they will use caves or rock crevices, thick undergrowth or even rabbit burrows.

The Mountain Brushtail Possum has a preference for short fat trees with few holes, but often shelters in dense clumps of epiphytic ferns. Termite mounds are sometimes used as den sites by Common Brushtail Possums living in northern Australia. The ceiling spaces of houses and sheds are also used by all brushtail possums living near human habitation

Scaly-tailed Possums shelter deep in rock piles. The Common Spotted Cuscus builds a small sleeping platform of twigs, but the Southern Common Cuscus dens in a tree hollow.

Feeding signs
Fruit scored by a brushtail possum's teeth is a sign well known to many householders. The damage is easily distinguished from that made by a bird's beak (Plate 68, page 200).

Other traces
All these possums leave claw marks on trees, and regularly used trees with fibrous bark show ample evidence of this (Plate 69, page 200).

RINGTAIL POSSUMS

Shelters

The Common Ringtail Possum builds a bulky nest, or drey, a roughly spherical ball about 25–30 cm across composed of twigs, strips of bark, leaves, ferns and grasses. There is an entrance hole 8–10 cm in diameter in one side of the nest, which is usually placed in a fork or among the branches of the tree or shrub where these grow close together (Plate 70, page 201). The nest is lined with moss, soft bark and leaves.

This ringtail also builds its nest in tree hollows, mistletoe clumps and in the roofs of houses and sheds.

The Green Ringtail does not use a den, but rests sitting on a branch, which it grips tightly. The other three species of rainforest ringtails use tree hollows when these are available, but also den in bulky clumps of epiphytes. The Herbert River Ringtail also builds a drey similar to that of the Common Ringtail in low, swampy rainforest and open forest when no holes are available.

Feeding signs

The young leaves of many eucalypt species are favoured by the Common Ringtail, and the tips of branches often show signs where these have been eaten. Eucalypt blossoms are also eaten, and chewed fragments of these may be found on the ground where the ringtails have been feeding. These fragments are usually small and can thus be distinguished from the larger debris left by birds, such as parrots, and the whole flower clusters dropped by flying-foxes (page 241). Garden flowers and fruit trees may also show signs where a ringtail has been feeding — torn petals and dropped fruit remains can sometimes be attributed to these possums.

The rainforest ringtails also mainly eat leaves, and sometimes fruits and blossoms. The leaves are usually bitten off at the petiole, a sign that can be used to locate the feeding places of these possums. Dropped remains of rainforest fruits, such as the Quandong and the Yellow Walnut, may be found, as these are sometimes eaten.

Other traces

The Common Ringtail often drops short lengths of wattle foliage when it is feeding in these trees, and they can be found scattered on the ground (Plate 71, page 201). Sometimes these are used to line the nest and they may be found some distance from wattle trees.

Ringtail possums, like brushtail possums, leave claw marks on trees and branches, but it is usually difficult to identify which possum is responsible without other evidence.

GLIDERS, STRIPED POSSUM AND LEADBEATER'S POSSUM

Shelters

All of these possums use tree hollows for their dens. The Greater Glider usually chooses a large, old tree, nesting in a hole high up in the trunk, which it sometimes lines with leaves and strips of bark.

Yellow-bellied Gliders, Sugar Gliders and Squirrel Gliders line their dens with leaves. Both living and dead trees are used by these gliders. Sugar Gliders have a preference for trees with numerous fissures (narrow cracks in the tree trunk), but they also den in hollow limbs (Plate 72, page 201).

Leadbeater's Possum uses a communal nest of shredded bark in the hollow centre of a large dead or living Mountain Ash, usually 10–30 metres above the ground. This possum has a preference for short, fat trees which have numerous holes in their trunks and a large quantity of dense vegetation surrounding them. (Plate 75, page 202).

The Striped Possum makes a leafy nest in a tree hollow, or amongst a clump of epiphytes.

Feeding signs

The Yellow-bellied Glider makes deep slashes in the bark of certain trees, so that it can lick the sap which exudes. The shape and depth of these gashes depends on the nature of the bark (Plate 77, page 203). The gashes can become quite large as the trunk or branch grows, and mature trees often show signs of sap sites made many years ago (Plate 78, page 203).

Pellets of bark which the Yellow-bellied Glider has chewed as it enlarges the gashes are ejected from the glider's mouth. These are roughly oval, and often folded along their length. They are a characteristic sign of this species, which also sometimes chews banksia flower-heads in a similar way (Plates 79, page 204).

Other species of gliders visit the sap sites made by the Yellow-bellied Glider, and Sugar and Squirrel Gliders make gashes of their own. On the trunks and branches of certain trees, Sugar Gliders gouge out the bark with their lower incisors, leaving vertical or horizontal gashes from which they lick the sap that exudes (Plates 73, 74, page 201).

These gliders, as well as Leadbeater's Possum, eat the clear gum that exudes from cracks and tears in the bark of some wattles (Plate 76, page 202). Leadbeater's Possum does not eat the darker gum that exudes from old borer holes.

The Striped Possum uses its long, sharp lower incisors to gouge away bark in its search for wood-boring grubs, leaving characteristic scars.

Other traces
Gliders use regular pathways through the forest, often landing on the same areas on the tree trunks along their way. Due to the friction and repeated claw marks made at these landing places, the bark is often rougher, or smoother, or a different colour to the rest of the tree trunk, depending on the nature of the bark.

PYGMY-POSSUMS, FEATHERTAIL GLIDER AND HONEY POSSUM

Shelters

The Eastern and Western Pygmy-possums both shelter in spherical nests of shredded bark or grass about 6 cm in diameter, built in small tree-holes or under loose bark on tree trunks. Nests have also been found in abandoned birds' nests and in hollow stumps (Plate 80, page 204). Western Pygmy-possum nests have also been found amongst the leaves of grass-trees.

Little Pygmy-possums make rough nests of strips of bark in old stumps, under ploughed turf, in wall cavities and a variety of other secluded places. They have also been found in small grassy nests in tussocks of Porcupine Grass.

The spherical nests of fern fronds and leaves of Long-tailed Pygmy-possums have been found in fern clumps and hollow stumps.

Nests of the Mountain Pygmy-possum have not been found in the wild, but in captivity they consist of a platform of grass over a shallow depression in the ground. This species probably makes similar nests below the rocks and boulders in the wild.

Feathertail Gliders construct globular nests of dry, overlapping eucalypt leaves. The nests are usually built in a tree hollow, but they have also been found in the dens of other possums and in such places as the boxes on telephone poles.

The Honey Possum does not build a nest, but shelters in the hollow stems of grass-trees and in abandoned birds' nests.

Feeding signs and other traces

The Mountain Pygmy-possum eats seeds of the Mountain Plumpine (*Podocarpus lawrencei*). It cracks a seed by holding it in one forepaw to the side of the mouth and biting with the large, sharp premolar. This cracks the seed transversely. Bush Rats also eat these seeds, but crack them lengthways (Figure 236; see also page 227).

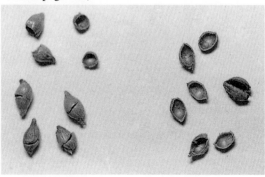

Figure 236 Seeds of Mountain Plumpine cracked transversely by a Mountain Pygmy-possum (left) and lengthways by a Bush Rat (right)

KOALA AND WOMBATS

Shelters

Koalas do not use any form of prepared nest but rest in the fork of a tree, usually in dense foliage. A thick fork which will provide protection from sun and rain may be used frequently by the same Koala. Although it may rest in the trees where it has been feeding, a Koala will also often use suitable forks in other trees and bushes.

All three species of wombats dig burrows. The soil or sand dug out of the burrow forms a large mound outside the entrance, making these shelters easy to find.

Common Wombat burrows vary in size and location. Typically, the oval or U-shaped entrance is about 40–50 cm across and 50 cm high, but old, well-established burrows may have entrances as much as a metre high. Frequently, some of the roots from nearby trees help to support the entrance.

The burrow is often dug into the side of a hill or a sloping bank. Short burrows, 1–2 metres in length, are commonly found in unfavourable places, such as the sides of dry creek beds or flat ground liable to flooding, and these are only used as temporary refuges. The major burrows are much deeper and longer, usually extending more than 10 metres underground. Burrows up to 30 metres long have been found (Plate 83, page 214).

Most Common Wombat burrows have only one entrance but some major burrows fork into two or more tunnels, and these may link up with others from adjacent burrows, forming a complex network of tunnels and entrances (Plate 82, page 213).

A Common Wombat usually has at least three major burrows that are regularly used, and more than one wombat may use the same burrow.

Southern Hairy-nosed Wombats construct extensive burrow systems, with clusters of entrances. There is usually a large central warren with smaller warrens surrounding it at a radius of about 100–150 metres. Entrances are often dug under shelving limestone rocks that support the entrances (Plate 84, page 214).

Most of the burrows of the small colony of the rare Northern Hairy-nosed Wombats are clustered in small groups. They are dug in deep river sand, the remains of the bed of a stream which dried up long ago. Most of the entrances are close to trees and the roots support the tunnel roofs (Plate 84, page 214).

Feeding signs

Cropped tussocks are often found where wombats have been feeding, but other evidence is needed to verify that wombats are responsible. In winter, Common Wombats living in the high country will scoop snow away in order to expose the plants, such as matrushes, that are underneath. Large holes in the snow are often found where these wombats have been feeding.

Other traces

The strong, sharp claws of the Koala leave characteristic marks on the trunks of smooth-barked trees. Koalas generally rely on the very sharp tips of their claws to hold their weight, and these tend to leave pock-marks on the smooth bark. If a Koala is in a hurry or loses its grip, longer rake marks are made (Plate 81, page 213).

All species of wombats use rubbing posts. Tree trunks, over-hanging branches, stumps or fallen logs may be used to rub the back or the flanks. This may leave a film of dust or moist earth, and often some hair, on the rubbing post; regularly visited posts may become smooth and polished (Plate 86, page 215).

Wombats make large depressions in sand or soft ground when they have a dust-bath. The wombat lies on its side, scratches the ground with its front claws and then scoops the loose sand or soil over its body. These wombat wallows are found at the bases of large trees, or on sandy tracks, creek banks and beaches.

The Common Wombat's habit of scoring the ground with its front paws as it deposits its scats leaves a characteristic V-shaped mark, which often remains for a long time after the scats have disintegrated (Plate 87, page 215).

Well-worn pathways are made by all species of wombats. These are particularly obvious near the warrens of the Southern Hairy-nosed Wombat (Plate 85, page 214).

*Plate 81
Characteristic
pock-marks made
by Koalas on the
smooth trunk of a
Southern Blue
Gum (Eucalyptus
globulus), with
some longer rake
marks also visible
(page 212).*

K. HANDASYDE

*Plate 82
A cluster of
Common
Wombat
burrows,
probably joined
underground
(page 211).*

213

A. HORSUP

Plate 83   The entrance to a major burrow of a Common Wombat (page 211).

Plate 84   Entrance to a burrow of the Northern Hairy-nosed Wombat (page 212).

Plate 85   A warren of the Southern Hairy-nosed Wombat (page 212).

J. STELMANN

214

Plate 86   Rubbing post of a Southern Hairy-nosed Wombat (page 212).

Plate 87   Scratch marks left by Common Wombat as it deposited the scats on the right (page 212).

K. ATKINSON
Plate 88   Nest of Long-nosed Bandicoot (page 221).

R. SOUTHGATE
Plate 89   Entrance to Bilby's burrow, in the Tanami Desert, WA (page 221).

Plate 90
A conical hole dug by Long-nosed Bandicoot with scats nearby (page 221).

Plate 91
Bilby diggings in the Tanami Desert, Western Australia (page 221).

R. SOUTHGATE

216

K. ATKINSON

M. OAKWOOD

Plate 92   Spot-tailed Quoll den in tree hollow (page 222).

Plate 93   Northern Quolls sometimes shelter in termite mounds (page 222).

Plate 94   Den of the Northern Quoll in rocky crevice. Note scats near entrance (page 222).

M. OAKWOOD

P. WOOLLEY

Plate 95   Entrance to the burrow of a Kowari (page 222).

W. LAIDLAW

Plate 96   The nest of a White-footed Dunnart, at the base of a tree (page 223).

P WOOLLEY

Plate 97   Entrance to the burrow of a Mulgara near Ayers Rock, Northern Territory. The burrow had one large hole and one pop-hole (positions arrowed) (page 222).

Plate 98   A small hole dug by a foraging Numbat (with a macropod scat next to it) (page 224).

Plate 99   A small crater and tracks show where a Marsupial Mole has dug its way into the sand (page 224).

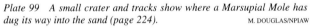

Plate 100   *Entrance to the nest of a Long-haired Rat (page 225).*

Plate 101   *Large untidy nest of a Black Rat, found in old machinery (page 225).*

Plate 102   *Nest of a Broad-toothed Rat, found in alpine grassland (page 225).*

BANDICOOTS AND BILBY
Shelters

Most species of bandicoots build nests of plant material, mainly leaves and grass, on the ground. They are usually constructed over shallow depressions, and are well concealed under forest litter or grass tussocks. The nest material is loose at each end, but there is often no obvious entrance hole, as the loose material is rearranged by the bandicoot as it enters and leaves the nest (Plate 88, page 216). The Long-nosed Bandicoot sometimes kicks soil over the top of its nest, possibly as waterproofing.

Hollow logs and grass tussocks are also used as shelters by bandicoots, and the Rufous Spiny Bandicoot, found on Cape York, is believed to also shelter in a burrow at times.

The Bilby constructs a burrow system, up to 3 metres long and 1.8 metres deep. The open entrance is usually at the base of a termite mound, spinifex tussock or small shrub (Plate 89, page 216).

Feeding signs

The characteristic conical pits dug by bandicoots as they search for invertebrate larvae in the soil have pointed bottoms. The size and shape of these diggings varies considerably. They may be as small as 3–4 cm across and 6–10 cm deep, but often they are more than 10 cm across and up to 15 cm deep. The nature of the soil affects the size and shape of the holes: holes in damp, loamy soil tend to be larger than those in firm soil (Plate 90, page 216). Often several holes are dug close together. Because bandicoots usually dig from one side of the hole, the loose soil is mostly piled around that side, but it is sometimes scattered in all directions.

The bandicoot's characteristic scats are often found near these holes (Plate 90, page 216).

Bilbies also dig conical holes, up to 10 cm deep, and areas where these animals are active are usually pock-marked with a number of holes. The soil is often scattered all around the hole (Plate 91, page 216).

QUOLLS, TASMANIAN DEVIL AND SOME SMALLER RELATIVES

Shelters

Both the Spot-tailed Quoll and the Northern Quoll often have their dens in hollow trees or hollow logs (Plate 92, page 217). They also shelter in caves, rocky crevices (Plate 94, page 217) and burrows, such as old goanna or rabbit burrows. Northern Quolls also use holes in termite mounds, particularly in spring, when mothers leave their young behind in these dens while they forage (Plate 93, page 217).

The Eastern and the Western Quolls generally shelter in grass or leaf-lined nests in burrows. These usually have a single entrance, about 10–20 cm high and wide. The burrows are often found at the base of a stump or under boulders. Old rabbit burrows are also used.

The Tasmanian Devil digs a burrow, often on sloping ground in the forest. These burrows may be up to 15 metres long, and have one or sometimes two entrances, about 20–30 cm in diameter. There may also be smaller entrances used by young cubs. Devils also shelter in caves and rock clefts, and often use active wombat burrows.

All quolls and devils living near human habitation will shelter under houses and sheds, and they all have a number of frequently used dens.

The Kowari digs burrows (Plate 95, page 218) and also uses those of other mammals, such as the Bilby and some rodents.

The Mulgara digs complex burrows with side tunnels and pop-holes (entrances to the tunnels) among the sand dunes of its desert habitat (Plate 97, page 218).

Feeding signs

Tasmanian Devils consume almost all of a carcass, including the skin and most of the bones. Menna Jones, studying quolls and devils in Tasmania, has found that only the heaviest bones are left, and these bones, such as wombat skulls and the pelvises of wombats, kangaroos and wallabies, usually have been chewed and pieces cracked off. (Devils dispose of large leg bones by holding them between their front paws and crunching them down like a carrot.) All that will remain of a recent carcass (up to a few months old) is the outline of fur on the ground and the remains of the gut contents. (Devils delicately remove the intestinal and stomach wall with their incisor teeth, leaving the contents [grass] behind).

PHASCOGALES AND ANTECHINUSES

Shelters

The Brush-tailed Phascogale makes nests lined with leaves or shredded bark in hollows in standing trees and stumps. The entrance holes to these nests are usually less than 10 cm in diameter. They sometimes shelter in the ball-shaped nests of babblers.

The Red-tailed Phascogale nests in hollow logs and hollow limbs of mature trees, particularly the Wandoo (*Eucalyptus wandoo*) and Rock Oak.

Most antechinuses construct a roughly spherical nest of dry plant material — grass and leaves — concealed in a protected place, such as a hollow log, small tree hollow, rock crevice, under a decaying log or in a grass tussock. Some also use the tunnels of Bush Rats and other rodents.

Feeding signs

The wings of insects such as grasshoppers, crickets and wasps are often found in and near the nests of phascogales and antechinuses.

DUNNARTS, PLANIGALES
AND OTHER SMALL DASYURIDS

Shelters

The Common Dunnart builds a cup-shaped nest of dry grass and leaves. These have been found in hollow logs, grass tussocks and grass-trees. The Fat-tailed Dunnart builds similar nests in hollow logs and under tussocks and rocks, and they have also been found under sheets of iron on the ground near farms. In arid areas this dunnart also shelters in deep cracks in the soil. Bark-lined nests of the White-footed Dunnart have been found on the ground (Plate 96, page 218), and in hollows in trees and logs.

The Kultarr, several other dunnart species and other small dasyurids live in arid country, where they shelter under saltbush and spinifex hummocks, in cracks and crevices in the soil and under rocks or logs. The Kultarr has also been found sheltering in the burrows of hopping-mice.

Feeding signs

Insect remains, such as wings, are often found in and near the nests of these dasyurids.

NUMBAT

Shelter
A nest lined with shredded bark in a hollow log is the usual shelter for the Numbat. A burrow 1–2 metres long is used in spring, when the mother leaves her unweaned young in a nest chamber at the end of the burrow while she forages.

Feeding signs
Numbats make small conical holes where they scratch away the leaf litter and stones covering shallow termite runways and ant nests (Plate 98, page 219).

MARSUPIAL MOLE

Shelter
The Marsupial Mole spends almost all its time buried in sand, clawing its way through the loose desert sand, which fills in behind it. After its occasional trips to the surface, the Marsupial Mole leaves a small hole as it re-enters the sand, but this is soon filled in by the shifting sand (Plate 99, page 219).

RATS AND MICE

Shelters

Many rodents shelter in burrows, in which they build nests of grass and other fibrous material.

The Bush Rat digs small burrows about 2–3 metres long in undergrowth, often at the base of a tree. The Swamp Rat digs a deeper burrow leading from its runway systems. The Dusky, Canefield and Long-haired Rats and Pale Field-rats all construct shallow burrows containing nests of shredded grass (Plate 100, page 220). All these rats also shelter in tussocks, tree roots and logs and other secluded places.

The introduced Black Rat builds its nest of any fibrous material it can find, such as grass, bark, shredded paper, rags, string and other material (Plate 101, page 220). It nests in tree hollows, roof spaces, wall cavities, old machinery, haystacks, poultry pens and similar places. It also digs shallow burrows.

The introduced Brown Rat digs deep burrows with entrance holes 6–10 cm in diameter; large spoil heaps from these can be found on the banks of creeks and drains, around barns and other farm buildings. They also use old rabbit burrows. Nests of bark, straw, rags or paper may be built under flooring, amongst crates and many similar places.

The introduced House Mouse sometimes builds complex, shallow burrows with several entrances. A spherical nest of grass, bark, paper, rags or other fibrous material is built in the centre of the burrow complex. These nests are also found in many places, such as house roofs, behind skirting boards, in machinery, wood piles, haystacks, or under sheets of corrugated iron and discarded timber in yards and paddocks. In this last case they are almost indistinguishable from the nests of the Fat-tailed Dunnart (page 223).

The Broad-toothed Rat builds large nests of shredded grass under logs and in dense grassy undergrowth (Plate 102, page 220).

Most of the 'false mice', or *Pseudomys* species, also dig burrows, and some also shelter in tussocks, in hollow logs, under pieces of bark and in the burrows of other animals, such as rabbits and lizards. Nests of grass or shredded bark are usually built in the burrows. In arid areas, the burrows are often more than half a metre under the surface (Plate 103, page 229).

The extensive burrows of the Silky Mouse have several entrances about 2 cm in diameter. Large heaps of sand mark the entrances in autumn when burrows are dug and extended.

The entrances of burrows of the New Holland Mouse are usually hidden under shrubs or dead leaves. Burrows up to 5 metres long have been found.

The Long-tailed Mouse builds its nest of fine bark strips

and other plant material under forest debris or in a small hollow in a log or stump.

The Western Pebble-mound Mouse of the arid Pilbara region makes piles of small stones which are penetrated by U-shaped tunnels connecting with pop-holes. Old mounds may be up to 9 square metres in area (Plate 104, page 229).

Stick-nest rats also build unusual nests. Although these rats are now extinct on the mainland — the Greater Stick-nest Rat is only found on an island off South Australia, and the Lesser Stick-nest Rat is totally extinct — old nests of both species have been found in breakaway caves and rock overhangs in several parts of central Australia. These large, communal nests were built with dead sticks and branches, woven firmly into structures as large as one metre high and two metres in diameter (Plate 105, page 222).

Rock-rats love among broken rocks, sheltering in crevices and ledges. The Black-footed Tree-rat shelters in tree hollows, and is sometimes found in the roofs of buildings; a nest of the rare Golden-backed Tree-rat, made of strips of pandanus leaves, was found in the foliage of a *Pandanus*, or screw-pine. The Brush-tailed Rabbit-rat also shelters in the leaves of the *Pandanus*, as well as in tree hollows.

Hopping-mice construct simple burrow systems, which have one or more broad, horizontal tunnels with vertical shafts opening to the surface. The pop-holes are about 3–4 cm in diameter. No loose soil or sand is found at the pop-holes, as the vertical shafts are dug from underneath. The original entrance hole is blocked up, but a pile of soil or sand is left outside it. This pile is not easy to find, as it soon weathers to a smooth surface.

The Giant White-tailed Rat has been found nesting in tree-hollows, but also probably nests in rock piles in some areas. The Fawn-footed Melomys make nests of plant material in trees; the Grassland Melomys weaves nests of grass around sugar cane leaves or grass stems, and also nests in trees.

The Water-rat usually shelters in a burrow dug into the side of a bank of a creek or other waterway. The opening is round, about 15 cm in diameter, and is often well protected by vegetation. Hollow logs are also used as resting places. False Water-rats have been found nesting in burrows in mangrove mudflats.

Feeding signs

The characteristic upper and lower incisors of rodents leave pairs of gnaw marks, and the size of these marks can be a clue to the size of the rodent responsible. The upper incisors, which hold the object, leave short, curved marks, while the lower incisors leave long furrows. Skins of firm berries, hard nut-shells and many thick-skinned fruits often show these tooth marks.

Places where grain is stored are often infested with the introduced rats and mice, and scatters of finely gnawed fragments of seed are evidence of their presence.

A rodent generally holds a seed between both forepaws and cracks the seed lengthways with the sharp incisors. In alpine country, the Bush Rat eats the seeds of *Podocarpus lawrencei* (Mountain Plum-pine) in this way, and the remains of these seeds can be distinguished from those eaten by the Mountain Pygmy-possum, which cracks them transversely (see page 210).

The Water-rat will take its food from the water to a feeding site to dismember it. Collections of empty shells, crab shells, fish bones and so on are found at 'dining tables' such as jetties and piers, moored boats and buoys, branches and stumps near creeks and other waterways and small clearings among reeds (Plate 106, page 230; also Figure 220, page 170).

Other traces

Rats and mice living in and near buildings often gnaw at manmade structures, and can cause considerable damage to electrical wiring, copper and plastic piping, window frames, doors and so on.

The above-ground runway systems used by rodents can be found passing under dense undergrowth or thick grass (see Figure 212, page 86). They often lead to the entrance to a burrow. Runways can also be seen in sandy country, where the small rodents run from one place of cover to another.

A characteristic greasy smear is left on regular pathways in buildings, in difficult passing places which rats constantly brush against as they pass, such as roof supports or corners of walls.

The presence of the House Mouse can often be detected by its characteristic stale, musty odour.

RABBIT AND HARE

Shelters

The rabbit shelters in a burrow, often as deep as half a metre below the surface. Single burrows are found, but usually there is a complex system of burrows, forming a warren. These have a number of entrances that are connected on the surface by well-worn paths or runs (Plate 107, page 230). Soil dug out of the burrow may form a heap outside the entrance, but some burrows are dug from the inside and these have no accumulated soil on the surface.

The hare does not dig a burrow, but shelters in a 'form' (a shallow depression in the grass) under small shrubs or among reeds or stones. Leaves, twigs and other debris are scraped out of the depression so that the hare lies on bare ground, except when young have been born in the form, when it will be lined with hair which the mother has pulled from her coat.

Feeding signs

Rabbits and hares often browse on young trees, biting off the shoots, some of which are often found lying on the ground near the tree. Because their teeth are very sharp, the bitten surfaces of the shoots are clean, oblique cuts, as though they have been sliced by a sharp knife (Plate 108, page 230).

Rabbits and hares also gnaw the bark of young trees, and can do much damage in plantations. Their tooth marks are distinctive: deep, usually longitudinal furrows.

Other traces

Rabbit scrapes are usually shallow scratchings, rounded at the base, with a small mound of soil or sand at one end. Often one or two of the round, pill-like scats are left on the mound (Figure 237). Rabbits also deposit their scats on slightly elevated places, where many scats will accumulate. These 'hills' are used as territorial markers (see Figure 221, page 171).

*Figure 237
A fresh rabbit
scrape, with
scats on the
mound.*

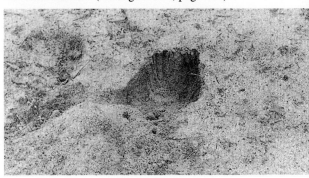

Plate 103   Entrance to the burrow of a Sandy Inland Mouse (page 225).

Plate 104   Nest of a Pebble-mound Mouse (page 226).

Plate 105   An old stick-nest rat nest in a rock overhang (page 226).

Plate 106   A Water-rat's feeding table at the end of a jetty (page 227).

Plate 107   A typical rabbit warren (page 228).

Plate 108   These clean, oblique cuts on branches of a cassia shrub are typical of rabbit damage (page 228).

D. WATTS/ANT

R. WELLS

Plate 109   A fox's den on the edge of pasture, with a few feathers nearby (page 238).

Plate 110   A tree regularly marked with urine by feral Dogs (page 237).

Plate 111   A decapitated egret, the remains of a fox's kill (page 238).

Plate 112   Damage caused by feral Pigs, where they have been rooting for food in the ground (page 239).

Plate 113   Damage caused by cattle stripping bark from a tree trunk (page 239).

232

Plate 114   Deep scratches made by a Sambar Deer rubbing its antlers on a tree (page 239).

Plate 115   This large mud puddle is the site of a buffalo wallow (page 239).

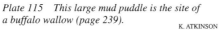
K. ATKINSON

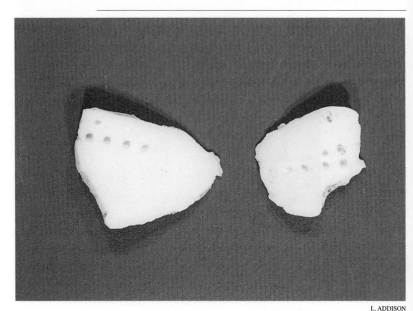

Plate 116   Pieces of cuttlebone, showing the holes made by a dolphin (page 241).

Plate 117   Flying-foxes spit out pellets of fibrous material after eating fruit (page 241).

Plate 118   Holes in gum nuts made by a parrot's beak (page 242).

Plate 119   She-oak cones crushed by Glossy Black-Cockatoos (page 242).

D.A. MEAGHER

Plate 120   Fan-shaped spread of sand at the entrance to the burrow of a White's Skink (page 242).

Plate 121   Ghost Crab hole (page 242).   K. ATKINSON

Plate 122   Yabbie hole in mud (page 242).

DOG, DINGO, RED FOX, CAT

Shelters

Dingoes and feral Dogs shelter in caves, hollow logs and in thick cover. They will also take over a wombat's burrow.

A fox usually digs an 'earth' or burrow, often in sandy or gravelly soil. The excavated soil is spread out in all directions from the entrance hole, unlike that of a wombat's burrow which forms a mound outside the entrance (Plate 109, page 231). Foxes also use wombat burrows and rabbit burrows, and sometimes shelter in hollow logs or the hollow limbs of trees up to about 4 metres above the ground.

Feral Cats also shelter in hollow logs and limbs of trees, but they can climb much higher than foxes. Feral Cat dens have also been found in termite mounds, in holes similar to those used by the Northern Quoll (see Plate 93, page 217).

Feeding signs

The prey of carnivorous mammals is not always found: small prey may be eaten whole, while larger prey is sometimes taken to the den or a secluded place. Traces of past meals — bones, feathers and so on — will often be found near the entrances of the dens of Dogs, Dingoes and foxes.

Figure 238
Dogs and foxes will sometimes cache food and dig it up later.

Foxes and Dogs will sometimes hide food by burying it or hiding it under bushes. These food caches are often found on beaches, where food such as beach-washed birds are buried for later consumption (Figure 238).

If the fresh remains of a carcass are found, it is sometimes possible to determine which kind of predator was responsible. Dogs, Dingoes and Red Foxes usually break the neck of their victim, then tear open the belly first to eat the liver and lungs of large prey, such as wallabies or lambs. However, a carcass will often soon be visited by other animals and birds, which also leave their signs, so positive identification of the original killer may be impossible.

Birds are usually decapitated and the head eaten whole; often the rest of the bird's carcass is left untouched (Plate 111, page 231), but wings torn off or tufts of feathers stuck together with saliva are further evidence that a Dog or Red Fox was the predator. In contrast, a bird of prey will pluck a bird it has killed, leaving a mass of feathers and down.

Cats generally take small prey and eat all of it, but they often leave evidence of their kill: the tail, and sometimes the hindquarters of a small possum or glider, or the wings of a bird.

Other signs

The characteristic musty smell of fox urine can often be detected near an occupied burrow and where the animal has recently passed by. Foxes, Dingoes and Dogs all use their urine for scent marking. Trees that are regularly marked with urine have a bare patch of ground at the base of the tree and a mossy-looking stain on the trunk directly above it (Plate 110, page 231).

Cats scratch out a small hole for their scats and urine, and then cover the hole carefully. These scratchings are very difficult to find in forest litter, but they can sometimes be detected on sandy beaches and in snow.

HOOFED MAMMALS

Shelters

The hoofed mammals do not construct any form of shelter. They lie down in any suitable place, choosing open or sheltered places as conditions suit them. Well-flattened grass with scats nearby will indicate these resting places. Most species have regularly used 'camps' in wooded country.

Feral Goats often rest under rock ledges and in caves. Feral Pigs have semi-permanent camps in dense undergrowth, usually near a swamp or stream. In hot weather, Pigs may also rest in dust wallows.

Feeding signs

Pigs leave areas of disturbed ground where they have been rooting for insect larvae, roots and fungi. Often these areas are large, and in some soils the marks made by the snout can be seen at the back of the holes (Plate 112, page 232).

Goats often browse on young trees and shrubs, stripping them of their bark and lower branches up to a height of 2 metres. Deer and cattle sometimes cause similar damage, and may reach to 3 metres above the ground (Plate 113, page 232).

Other traces

Deer rub their antlers on trees and bushes, which causes fibrous bark to fray and scratches smooth bark (Plate 114, page 233). Male deer cast their antlers once a year, replacing them with progressively larger ones. The cast antlers are usually found singly, and most are easy to identify, as the antlers of each deer species have a characteristic shape (Figure 239, page 240).

Buffalo, Pigs and deer all disturb large areas of ground when they take mud baths. These wallows are usually found in swampy areas and along watercourses (Plate 115, page 233).

Often near a wallow there are trees with muddy smears where the animals have rubbed themselves after their mud baths.

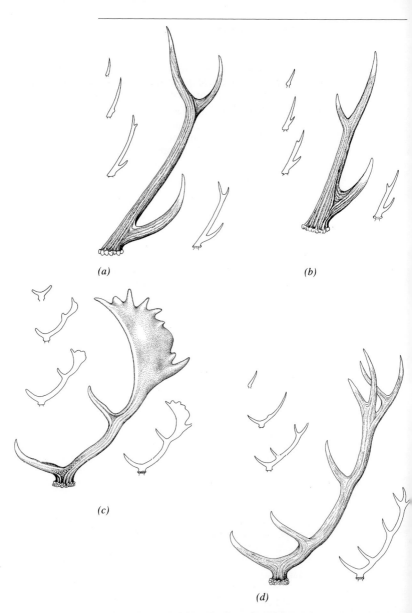

Figure 239   *Antlers of four deer species. Each figure shows the fourth year's antler, reduced by about 10 times, with the first, second, third, and fourth year's antlers also shown greatly reduced. (a) Red Deer, (b) Sambar Deer, (c) Fallow Deer, (d) Hog Deer. (T. Wright)*

MARINE MAMMALS

Shelters
None of the marine mammals have a permanent shelter. Those that come ashore to breed and rest do so at suitable areas along the coastlines they inhabit. They may use crevices and gullies in rocky areas as shelter, but they rarely move far from the shore.

Feeding signs
As these mammals feed at sea, feeding signs are not generally found, but a sign of a dolphin's meal is sometimes washed up on ocean beaches: cuttlebone — the internal 'shell' of cuttle-fish or squid — with a row of holes made by the even, conical teeth of the dolphin (Plate 116, page 234).

BATS

Shelters
Bats do not construct any form of shelter, but make use of natural, or sometimes artificial, cover.

The megabats generally roost in dense vegetation. Flying-foxes roost in the branches of large trees in forests or man-groves. Fruit-bats rest in dense vegetation, but have also been found in old mines, abandoned houses and occasionally in caves.

The Ghost Bat roosts in large caves, deep rock fissures and in old mines.

Many of the microbats also favour caves, mines, rock fissures and buildings. For some, such as the horseshoe-bats, the caves must be warm and humid. Other roost sites of the many species of small bats include tree-hollows and spouts, old birds' nests, under sheets of bark, and in rolled-up canvas awnings.

Feeding signs
Flying-foxes and other fruit-eating bats take a mouthful of fruit, press the pulp against the roof of the mouth to extract the juice, and then spit out the fibrous material. These 'spits' can be found under the trees where these bats have been feeding (Plate 117, page 234). Damaged fruit on trees often shows the imprint of the megabat's two rows of sharp teeth.

When feeding on eucalypt blossom, flying-foxes often drop whole flower clusters onto the ground below.

The Ghost Bat, the only carnivorous bat in Australia, feeds in caves and under rock overhangs, where it drops the remains of its prey, such as frogs, lizards, small mammals and birds. These remains accumulate under the feeding sites.

The microbats are all insectivorous. Large collections of insect wings and other discarded fragments are found in caves and other places where these bats roost.

OTHER ANIMALS

The vast array of shelters and signs of birds, reptiles, amphibians and insects are beyond the scope of this book, but a few that are often encountered are mentioned below, as these may be mistaken for similar signs of mammals.

BIRDS

Shelters Some bulky nests built in tree forks or branches can be mistaken for the dreys of the Common Ringtail Possum. A lyrebird's nest of sticks, leaves and fern fronds is usually built on the ground or on a sloping bank, but it is sometimes built in the head of a tree-fern or up to 25 metres high in a tree fork.

Small birds' nests built in tussocks and shrubs are similar to the nests of antechinuses and some rats and mice.

Many bird species nest in the same kinds of holes in tree trunks and limbs as those used by many mammals for their shelters.

Feeding signs The regurgitated pellets of birds are similar to some mammal scats. They are described and illustrated in the Scats section (pages 178–182).

On fruit and seeds it is usually easy to distinguish between the beak marks of birds and the tooth marks of mammals, but some feeding signs are similar. For example, the strong beaks of parrots leave a hole in a gum nut that is similar to that made by the sharp teeth of a large rat (Plate 118, page 235), and the crushed remains of she-oak cones, a feeding sign of the Glossy Black-Cockatoo, could be mistaken for the work of possums (Plate 119, page 235).

Eucalypt blossoms are visited by many bird species, such as rosellas and lorikeets, and florets are often found under the trees. Possums and gliders leave similar feeding signs, but flying-foxes drop whole eucalypt flower-heads to the ground (see page 240).

The scratchings of birds can be mistaken for similar mammal scratchings. The strong feet of lyrebirds, for instance, scatter leaf litter and soil in all directions, and some birds scratch out shallow hollows in the ground in their search for food.

REPTILES

Burrows dug by lizards are usually oval, with a fan-shaped spread of soil or sand around the entrances, not a mound as is the case with mammal burrows (Plate 120, page 236). Many reptiles shelter in hollow logs and tree-holes, sometimes sharing these with resident mammals. Goannas also nest in termite mounds.

INVERTEBRATES

Leaf damage by caterpillars and other insect larvae could be mistaken for that caused by ringtail possums and other foliage-eating mammals. Crabs and other invertebrates shelter in holes in the sand or mud (Plates 121, 122, page 236).

SKULLS, LOWER JAWS AND OTHER BONES

When animals die or are killed in the wild, their bodies are quickly disposed of by predators and scavengers. Foxes, dingoes and feral dogs, feral cats, ravens, eagles and other birds of prey, reptiles such as goannas, as well as a host of invertebrates, will consume an animal's flesh and viscera, but the bones are extremely long-lasting, and these provide valuable information.

With larger mammals, the skull, lower jaw, the long bones of the front and back legs and some of the backbone are usually left. These, cleaned up by beetles and other insects and whitened by sun and rain, are conspicuous in the bush, although they may have been widely scattered by the scavengers. Skulls and bones may also be found on beaches, as the remains of seals and other marine mammals are sometimes washed up on the shore. Bird skeletons are also often found on beaches, and it is useful to be able to distinguish between mammal and bird remains (see page 329).

Skeletal remains are sometimes found at the 'kill', or outside the predator's den. When smaller animals are eaten, many of their bones are swallowed, and these bones (or fragments of them) may be found in the scats of carnivorous mammals. They are also found in the regurgitated pellets of some birds. Owls often swallow small prey whole, and their digestive juices have little effect on bones, so their pellets provide an excellent means of identifying some small mammals. Fragments of larger prey, such as ringtail possums and brushtail possums, are also often identifiable (Figures 240, 241). Pellets are mainly found on the ground below the birds' nests and roosting sites (Plate 56, page 155).

The pellets of eagles, hawks and other birds of prey usually contain hair and feathers but few identifiable bones. Most of these birds often take very small prey, including birds and lizards, and their strong digestive juices dissolve much of the bony material. (See Figures 229, 230, pages 180, 181).

Although it is relatively easy to find the bones of an animal in the wild, it is more difficult to identify them accurately. The skulls and teeth are the most distinctive parts of an animal's skeleton. While only an expert can differentiate between closely

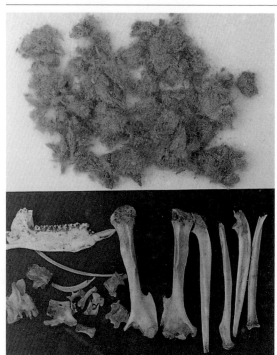

*Figure 240
Contents of
Powerful Owl
pellet*

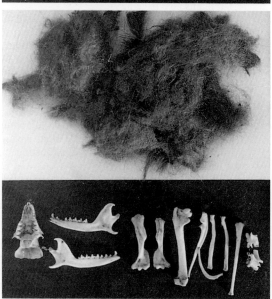

*Figure 241
Contents of
Sooty Owl
pellet*

244

related species, the Guide that follows should enable you to determine the group to which an animal belongs by examining its skull or lower jaw, and, in some cases, its humerus (upper arm bone) or femur (thigh bone). You will probably be able to identify a specimen as a wallaby, or a possum, or a rat, for example, but because only a few species are illustrated in this Guide, it may not be possible to identify which particular species of wallaby, possum, or rat you have found. If you want positive identification it is best to send or take the bones to an expert at the museum in your state capital city, or at the state wildlife authority, such as the National Parks and Wildlife Service in New South Wales.

The species illustrated in the Guide are, wherever possible, the ones most likely to be found and they are typical of the group to which they belong. The Eastern Grey Kangaroo, for example, is illustrated because it has a broad distribution and has many similar features to the other large kangaroos. In some cases, more than one representative of the group is shown. For example, two bandicoots are illustrated (one from each of the genera *Perameles* and *Isoodon*, represented here by the Long-nosed Bandicoot and the Northern Brown Bandicoot), because it is relatively easy to distinguish between them. By referring to the distribution maps in the Scats section it may even be possible to make a positive identification to species level.

Figure 242
Skeleton of
wombat.

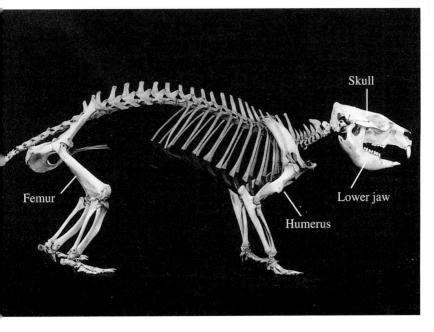

Skull

Femur

Lower jaw

Humerus

245

What to do when you find a skull or bone

1   Decide the anatomical identity of the specimen you have found, i.e. what part of the skeleton it comes from. If it is a complete skull or lower jaw, or a humerus or femur, this will be not be difficult. (Refer to the examples on the opposite page.)

   If the specimen is broken, it is still often possible to determine that it is part of the skull, or a fragment of a lower jaw, and so on.

   If it is a bone such as a tibia, scapula, pelvis, rib or vertebra, for example, you will not be able to identify the animal using this Guide.

2   Measure the maximum length of the specimen, *e.g.* the length of the skull from the front teeth to its base, as shown on the opposite page. If it is broken, try to assess how long the whole specimen would have been. The specimens shown in the Guide are arranged according to size, and they are from adult animals except where otherwise indicated. Difficulties will arise if you are trying to identify immature bones. There is also a problem in some groups in which there is a size difference between male and female. This is particularly noticeable in the kangaroos and wallabies, where it makes identification of the humerus and femur very difficult because a female kangaroo's long bones are about the same size as those of a large male wallaby.

3   Use the measurement you have made and the length key on page 251 to find the appropriate page or pages in the Guide.

Skull

For most of the species illustrated, the Guide shows the skull from underneath. Some are shown in profile as well as from underneath. The horse and domestic cattle are only shown in profile. All skulls in the Guide are shown actual size, except where stated otherwise.

   Turn your specimen so that it is in the same position as the illustrated skulls. Compare the shape of the skull, the number of teeth or sockets present (some teeth may be missing) and the spacing of the teeth. Examine the molars and the patterns of their cusps and compare them with the examples in Figure 244.

   In some cases, the bullae or ear bones (the rounded bony projections that enclose the two middle ears) have a characteristic size and/or shape. These and other distinctive characteristics are noted on the page facing each illustration. The dental formula is also given on the facing page, showing the number

*Figure 243
Skull, lower jaw,
humerus, femur.*

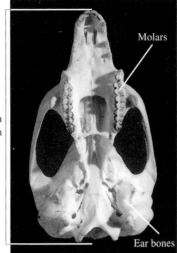

Molars

Maximum
length

Skull from underneath

Ear bones

*Lower jaw —
profile of right
jaw, inner
(tongue) side*

Maximum
length

The bending of
this area to form
a 'shelf' is a
characteristic of
most marsupials

Rough area
where jaws were
joined in life

*Right humerus,
from 'front' of
animal*

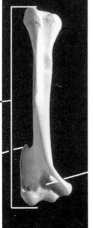

Maximum
length

Flange
with
point

*Right femur
from 'rear' of
animal*

Prominent
knob, a
characteristic
of the femur

Maximum
length

Hole near
base, a
characteristic
of the
humerus

of teeth present on each side of the jaw in a living adult:

I = incisor, C = canine, P = premolar, M = molar.

Some of these teeth may be lost after death, and some kangaroos lose teeth in life (see page 256). The premolars and molars together are sometimes called the cheek teeth.

Lower jaw
For most of the species illustrated, the Guide shows the profile of the right lower jaw from the inner (tongue) side. In cases where the lower jaws are usually found joined (wombats and Koala) they are shown from above. The lower jaws of the horse, domestic cattle, and Short-beaked Echidna and Platypus are shown on the same illustration as their skulls, in profile from the outer (cheek) side. All lower jaws in the Guide are shown actual size, except where stated otherwise.

Hold your specimen so that it is in the same position as the illustrated lower jaw. Compare the number of teeth or sockets present (some teeth may be missing) and the spacing of the teeth. Examine the molars and the patterns of their cusps and compare them with the examples on page 249.

Distinctive characteristics are noted on the page facing each illustration. The dental formula is also given on the facing page, showing the number of teeth present on each side of the jaw in a living adult.

Humerus
The right humerus is shown in all illustrations in the Guide. It is shown from what would be the 'front' of the animal if the bone was in its natural position, and photographed resting on a flat surface. All humeri in the Guide are shown actual size, except where stated otherwise.

Hold your specimen so that it is in the same position as the illustrated humerus. Distinctive characteristics are noted on the page facing each illustration. (Note: the point on the flange, which is a common characteristic in most marsupials, may be rounded in a weathered specimen.)

Figure 244
Examples of
cheek teeth

High-crowned, with sharp crescent-shaped ridges. (Typical of horse, cow, sheep, goat.)

Low-crowned, with four cone-shaped cusps. (Typical of bettongs, rat-kangaroos, brushtail possums, some small possums and gliders, pygmy-possums.)

High-crowned, rectangular molars with pairs of transverse ridges separated by a deep trough which is crossed by a longitudinal ridge. (Typical of kangaroos and wallabies.)

Molars triangular or squarish in plan, with several sharp cusps. (Typical of bandicoots.)

Two rows of crescent-shaped cusps. (Typical of foliage eaters, such as Koala, ringtail possums, Greater Glider.)

Strong shearing teeth with sharp cusps. (Typical of carnivores that do not chew their food, such as Dog, Red Fox, Cat.)

Strong molars, triangular in plan with several sharp cusps. (Typical of carnivores that chop up their food before swallowing, such as quolls, Tasmanian Devil. Also small insectivores, such as antechinuses, dunnarts.)

Low-crowned, with flat wear faces, no cusps. (Typical of wombats.)

Complex patterns of cusps and ridges. (Typical of all rats and mice.)

Femur

The right femur is shown in all illustrations in the Guide. It is shown from what would be the 'rear' of the animal if the bone was in its natural position, and photographed resting on a flat surface. All femurs in the Guide are shown actual size, except where stated otherwise.

Hold your specimen so that it is in the same position as the illustrated femur. Distinctive characteristics are shown on the page facing each illustration.

Scale bars

Scale bars are provided on all photographs of bones. Small dimensions of the bars are millimetres.

All skull, jaw and bone photographs in the Guide were taken by Carl Bento of the Australian Museum, Sydney.

Key to Skulls, Lower Jaws, Humeri and Femurs

SKULL	Illustrations	Pages
Length		
30 cm or more	S1	252–253
20–30 cm	S2, S3, S7	254–7, 264–5
13–20 cm	S3, S4, S5, S6, S7, S8	256–267
10–13 cm	S9, S10, S11, S12	268–275
7–10 cm	S10, S11, S12, S13, S14, S15	270–281
5–7 cm	S15	280–1
3–5 cm	S16, S17	282–5
1–3 cm	S17	284–5

LOWER JAW		
Length		
30 cm or more	S1, D1	252–3, 286–7
14–30 cm	D1, D2	286–9
9–14 cm	D3, D4, D5	290–5
8–10 cm	D6, D7, D8	296–301
6–8 cm	D8, D9	300–303
4–6 cm	D10, S10	304–5, 270–1
3–4 cm	D11, D12	306–9
1–3 cm	D12	308–9

HUMERUS		
Length		
11–18 cm	H1, H2	310–3
7–11 cm	H2, H3	312–5
4–7 cm	H3, H4	314–7
1–4 cm	H4, H5	316–9

FEMUR		
Length		
16–30 cm	F1	320–1
10–16 cm	F2, F3	322–5
7–10 cm	F3, F4	324–7
1–7 cm	F4, F5	326–9

Guide to Skulls, Lower Jaws, Humeri and Femurs

A **Horse**

1   Distance from eye socket to front of jaw longer in horse than in cattle, sheep, goats or deer.
2   Large incisors in both upper and lower jaws.
3   Small canine usually present in males, not in females.
4   Long gap between incisors and cheek teeth.
5   Large premolars and molars have crescent-shaped ridges, similar to cattle (B), and sheep and Goat (S2).

Dental formula:
Upper jaw — I3, C0–1, P3–4, M3
Lower jaw — I3, C0–1, P3, M3
Similar species: Donkey

B **Domestic Cattle**

6   If horns are present they project towards the sides of the head.
7   No upper incisors. Lower incisors bite against a hard pad in upper jaw.
8   Sockets where lower incisors were present. These are often lost after death.
9   Large premolars and molars have crescent-shaped ridges along their grinding surfaces, similar to Horse (A), sheep and Goat (S2).
10 Cheek teeth are often lost after death.

Dental formula:
Upper jaw — I0, C0, P3, M3
Lower jaw — I3, C1, P3, M3
Similar species: Water Buffalo, some species of deer

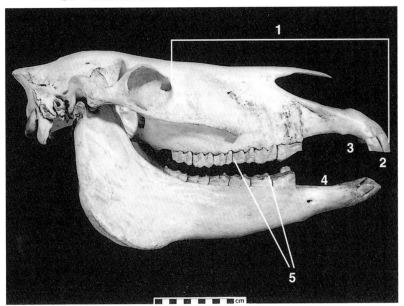

A **Horse**

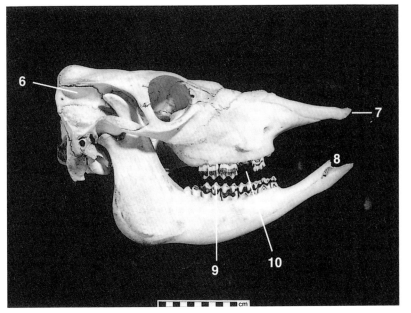

B **Domestic Cattle**

253

Note that the photographs have been reduced to 1/3 natural size.

A **Pig**

1   This skull is from a young animal. The canines are small and not all the cheek teeth have erupted. The canine grows at an angle to the jaw, and in older males it may form a prominent tusk.
2   The cheek teeth have many knobbed cusps.

Dental formula: I3, C1, P4, M3
Similar species: None

B **Sheep**

3   No upper incisors.
4   Cheek teeth have crescent-shaped ridges.
5   If horns are present they project towards the sides of the head.
6   This area is more pointed in sheep (B) than in Goat (C).

Dental formula: I0, C0, P3, M3
Similar species: Goat, some species of deer

C **Goat**

7   No upper incisors.
8   Cheek teeth have crescent-shaped ridges.
9   If horns are present they project towards the back of the head. (Not present in this specimen).

Dental formula: I0, C0, P3, M3
Similar species: Sheep, some species of deer

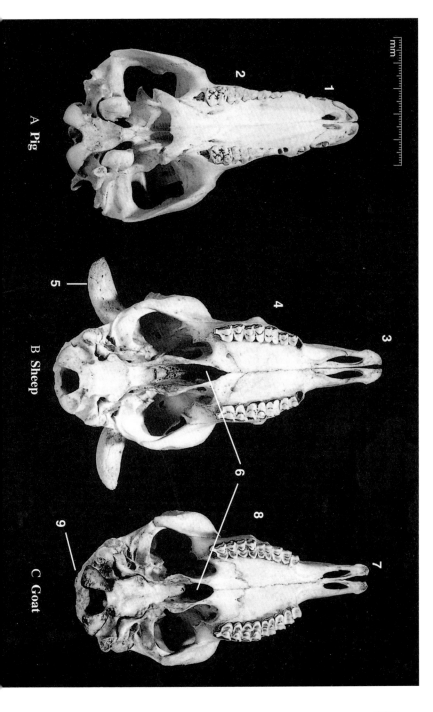

A Pig

B Sheep

C Goat

255

Note that the photograph has been slightly reduced in size (about 4/5 natural size).

Eastern Grey Kangaroo

1 Three pairs of upper incisors (also see S6). All kangaroos and wallabies have three pairs.

2 Long gap between incisors and cheek teeth — characteristic of grazing animals.

3 Premolar is shorter than the second molar.

4 The cusps on each molar form a pair of sharp transverse ridges. This is a characteristic of all kangaroos and wallabies.

5 The cheek teeth in kangaroos move forward in the jaw as the animal grows; they become worn and fall out in front, one at a time, so the number present varies with age. A very old animal might have only one or two worn molars left in each jaw. This kangaroo had lost the premolar on one side and would soon have lost the other one. All four molars have erupted. It was probably about five years old.

6 There are no large holes at the rear of the palate. These holes are also absent in the Western Grey Kangaroo. Several small holes may be present in older skulls.

Dental formula: I3, C0, P1, M4
Similar species: Western Grey Kangaroo, Red Kangaroo, Common Wallaroo, Antilopine Wallaroo, Black Wallaroo. Also Large Wallabies — see Swamp Wallaby (S5), and Red-necked Wallaby (S4).

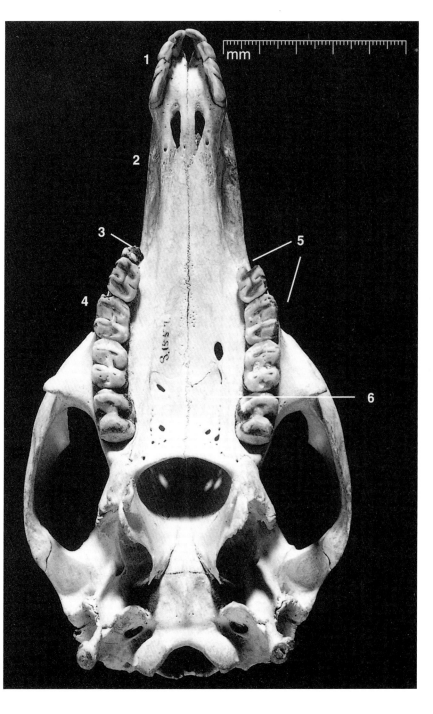

Red-necked Wallaby

1   Three pairs of upper incisors (see also S6).
2   Long gap between incisors and cheek teeth.
3   Premolar is smaller in Red-necked Wallaby than in Swamp Wallaby (S6).
4   Molar cusps form pairs of transverse ridges.
5   These holes at the rear of the palate are usually smaller in the Red-necked Wallaby than in the Swamp Wallaby (S5).

Dental formula: I3, C0, P1, M4
Similar species: Agile Wallaby, Black-striped Wallaby, Whiptail Wallaby, Western Brush Wallaby, Tammar Wallaby, Swamp Wallaby (S5). Parma Wallaby skull is similar but smaller. Skulls of large male wallabies may be similar in size to skulls of female kangaroos.

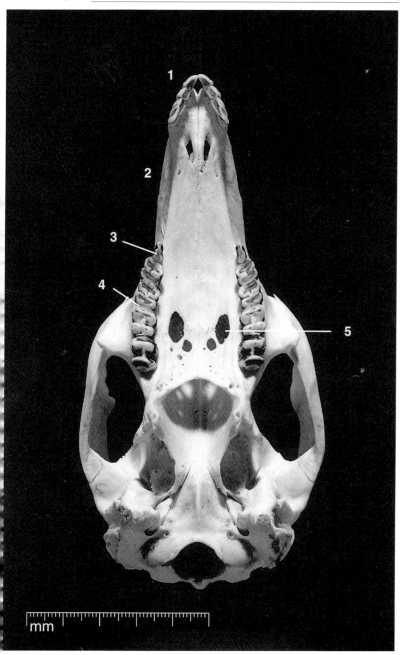

Swamp Wallaby

1 Three pairs of upper incisors (see also S6).
2 Long gap between incisors and cheek teeth.
3 Premolar is a strong, ribbed, cutting tooth. It is longer than the longest molar. This tooth is larger in the Swamp Wallaby than in other large wallabies.
4 Cusps on molars form pairs of transverse ridges.
5 These holes at the rear of the palate are usually larger in the Swamp Wallaby than they are in the Red-necked Wallaby (S4).

Dental formula: I3, C0, P1, M4
Similar species: Red-necked Wallaby (S4), Agile Wallaby, Black-striped Wallaby, Whiptail Wallaby, Western Brush Wallaby, Tammar Wallaby. Parma Wallaby skull is similar but smaller. Skulls of large male wallabies may be similar in size to skulls of female kangaroos.

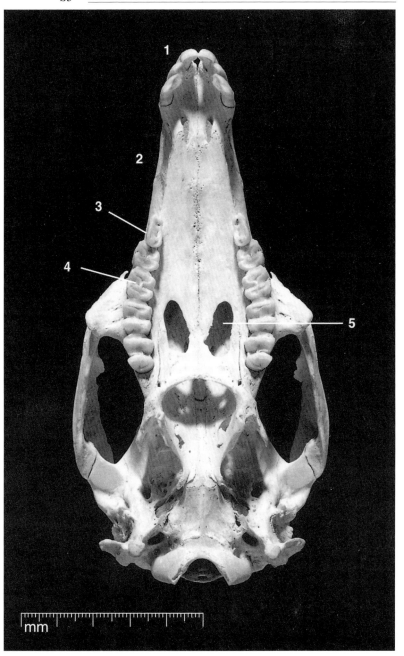

A Eastern Grey Kangaroo

1   The third upper incisor has a notch near its centre. It is a long tooth, generally longer than the combined lengths of the other two incisors. In older animals this third incisor may be shorter, and the notch may be worn away.

B Red-necked Wallaby

2   The third upper incisor has a strongly marked notch at about the middle of the tooth. Other large wallabies in the *Macropus* genus also have this notch to varying degrees. This third incisor is about the same length as the combined length of the other two incisors.

C Swamp Wallaby

3   The notch in the third incisor is less defined and is about two-thirds of the way along the tooth. This third incisor is generally smaller than in the other large wallabies — usually slightly shorter than the combined length of the other two incisors.

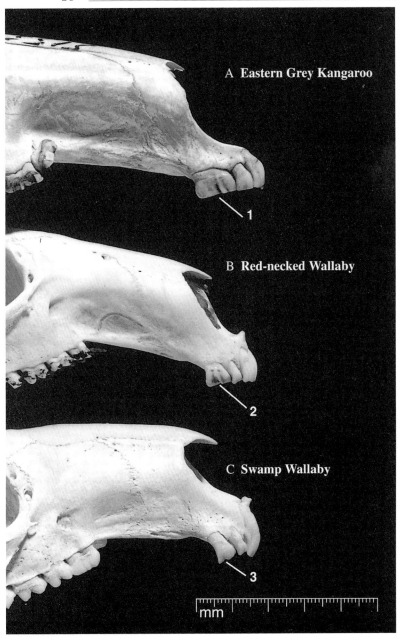

A Eastern Grey Kangaroo

B Red-necked Wallaby

C Swamp Wallaby

A **Australian Fur-seal**

This is the skull of a female; the skull is much larger in the male seal and the canines are larger.

1   The third incisor is larger than the others.
2   The canine is a strong curved tooth.
3   The cheek teeth are small and peg-like. There is no difference between the molars and premolars. The seal's soft-bodied prey is swallowed without chewing. Compare with Dog (B).
4   Large rounded brain case.

Dental formula: I3, C1, P&M 6
Similar species: New Zealand Fur-seal, Australian Sea-lion, Leopard Seal, Elephant Seal

B **Dog/Dingo**

This is the skull of a Dingo; feral Dogs of many mixed breeds will have skulls with a variety of shapes and sizes.

5   Third incisor is larger than the others, and a similar shape to the canine.
6   Canine is strong and large.
7   These holes in the front of the palate are level with or just behind the canines. In the fox they are further forward in the palate (see photograph opposite).
8   Largest cheek tooth is the fourth premolar.
9   Projection over the eye socket is at a sharper angle in the fox; otherwise the skulls are very similar.

Dental formula: I3, C1, P4, M2
Similar species: Red Fox, Tasmanian Devil (S9), Thylacine. The Cat's skull is much broader and has a shorter snout, larger ear bones, and two premolars (not four as in Dog) (see photograph opposite).

Fox and Cat skulls (half natural size).

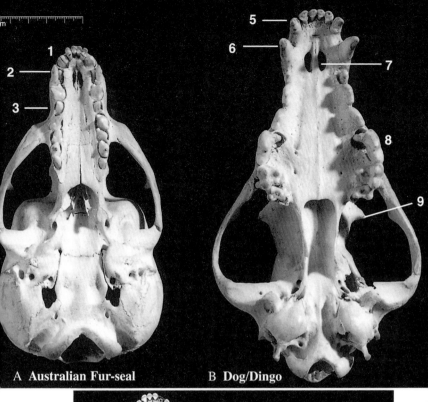

A **Australian Fur-seal**   B **Dog/Dingo**

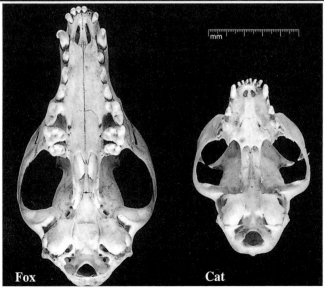

Fox   Cat

Common Wombat

1 One pair of upper incisors, with a sharp edge on a flattened wear face.

2 Large gap between incisors and cheek teeth.

3 Large cheek teeth also have flat wear faces. The molars have a characteristic double column structure and are curved (see below). Crowns have a hard outer ridge of enamel enclosing the softer dentine.

4 Skull is broad and strong, and flattened on the upper surface.

Dental formula: I1, C0, P1, M4
Similar species: Southern Hairy-nosed Wombat, Northern Hairy-nosed Wombat

The wombat's teeth have open roots and continue to grow throughout life. They often fall out after death.

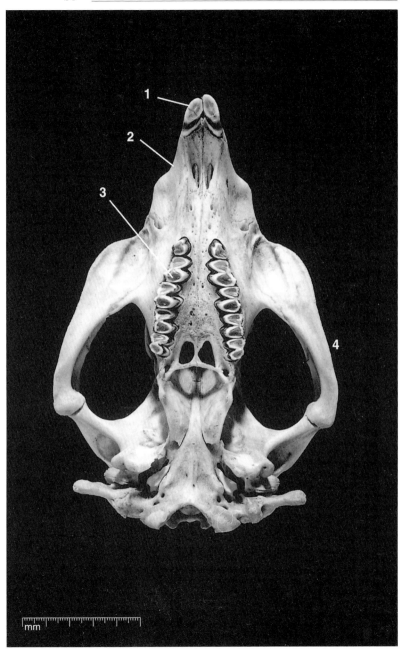

A **Koala**

1 Three pairs of upper incisors, first pair large and chisel-shaped, others small.
2 Small peg-like canine.
3 Premolars are strong cutting teeth.
4 Molars each have four prominent, crescentic cusps. These become worn and flattened with age.
5 Skull has an unusual oblong, parallel-sided shape.
6 Very large, prominent ear bones.

Dental formula: I3, C1, P1, M4
Similar species: none

B **Tasmanian Devil**

7 Four pairs of upper incisors.
8 Hollow where point of lower canine fits when mouth is closed.
9 Very large, strong canine.
10 Two crowded premolars.
11 Molars have sharp cusps.
12 Fourth molar is a small ridge at right angles to other molars.
13 Prominent projection over eye-socket does not show in this photograph.
14 Skull is broad, almost triangular in shape, with very small brain case.

Dental formula: I4, C1, P2, M4
Similar species: Dog/Dingo (S7), fox, Thylacine

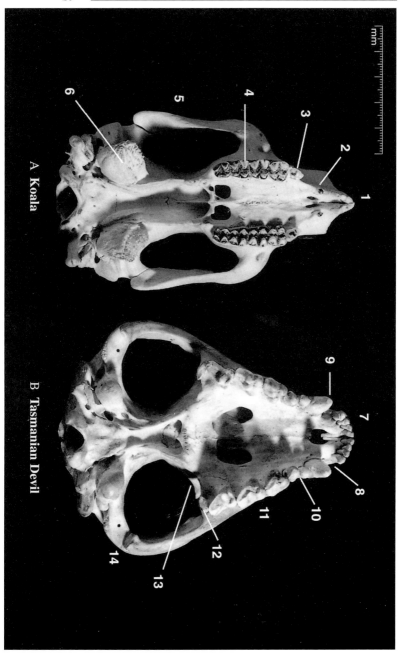

A Koala

B Tasmanian Devil

Note that the photographs of the skulls and lower jaws in profile have been reduced to 1/2 natural size.

A **Short-beaked Echidna**

1   Large, bulbous brain case.
2   Lower jaw is a long, thin, curved bone.
3   Front part of upper jaw is joined (this may be broken after death).
4   Echidna has no teeth. Food is pressed against the roof of the mouth by the tongue.
5   The long snout, or rostrum, is used to probe into ants' nests.

Similar species: none. Skull can be mistaken for a bird's skull.

The bird's skull is more fragile than the echidna's, and seen from the side the brain case is smaller and less bulbous.

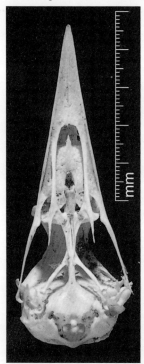

B **Platypus**

6   Lower jaw is stronger than in the echidna (A).
7   Front part of upper jaw is not joined. These bones form the wide, flattened snout — the 'duck-bill'.
8   Platypus is also toothless. Horny molar plates grind the food.

Similar species: none.

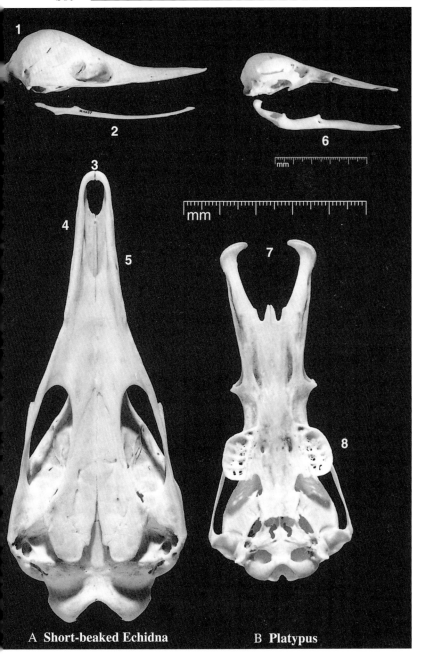

A **Short-beaked Echidna**   B **Platypus**

A Red-necked Pademelon

1   Three pairs of upper incisors, as in all wallabies and kanga-roos (also see S12).
2   Long gap between incisors and cheek teeth.
3   Premolar is a long, ribbed cutting tooth (also see S12).
4   Molars cusps form pairs of transverse ridges, characteristic of kangaroos and wallabies.
5   Relatively large holes at rear of palate — compare with kangaroo and large wallabies (S 3–5).

Dental formula: I3, C0, P1, M4
Similar species: Red-legged Pademelon, Tasmanian Pademelon, Brush-tailed Rock-wallaby (S12) and other rock-wallabies, Northern Nailtail Wallaby (S12) and Bridled Nailtail Wallaby, Lumholtz's Tree-kangaroo (S12) and Bennett's Tree-kangaroo, Spectacled Hare-wallaby (S12), Quokka

B Spot-tailed Quoll

6   Four pairs of small upper incisors. These may be lost after death.
7   Hollow where point of lower canine fits when mouth is closed.
8   Large, strong canine.
9   Premolars less crowded than in Tasmanian Devil (S9).
10 First three molars triangular in shape, with sharp cusps.
11 Fourth molar is a small ridge at right angles to other mo-lars.

Dental formula: I4, C1, P2, M4
Similar species: Eastern Quoll, Northern Quoll, Western Quoll, Tasmanian Devil (S9)

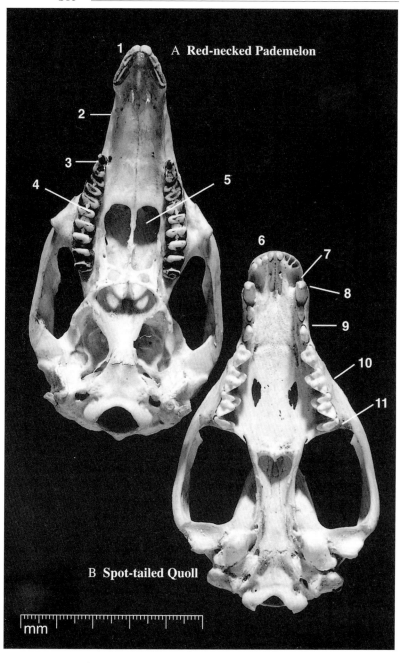

A Red-necked Pademelon

B Spot-tailed Quoll

273

A Red-necked Pademelon

1 Conspicuous notch near the end of the third incisor (may be worn away in older animals).
2 Premolar is a strong, ribbed cutting tooth. It is shorter than the longest molar.

B Brush-tailed Rock-wallaby

3 Conspicuous notch about two-thirds of the way along the third incisor.
4 Long, straight gap between incisors and cheek teeth.
5 Premolar is long and blade-like, with several grooves.

C Northern Nailtail Wallaby

6 Incisors are long and narrow, all about the same width.
7 A small canine may sometimes be present in this position and is an important diagnostic feature.
8 Premolar is very small.

D Lumholtz's Tree-kangaroo

9 Small third incisor has small notch near its centre.
10 A small canine is present. This canine is an important diagnostic feature.
11 Premolar is blade-like and ribbed, and is longer than the largest molar.

E Spectacled Hare-wallaby

12 Robust incisors, all about the same width.
13 A small canine is present. This canine is an important diagnostic feature.
14 Premolar is small, and shorter than the largest molar.

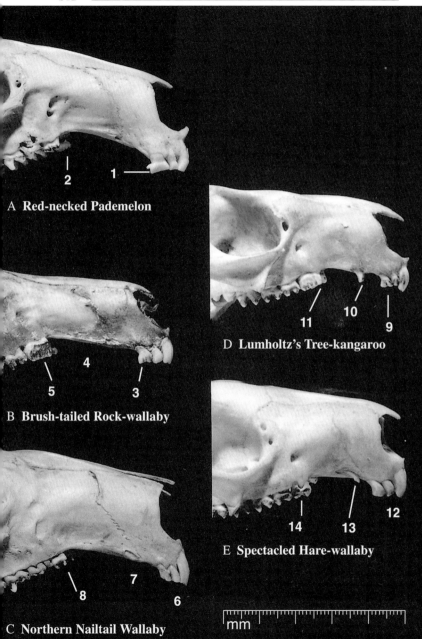

A Red-necked Pademelon

B Brush-tailed Rock-wallaby

C Northern Nailtail Wallaby

D Lumholtz's Tree-kangaroo

E Spectacled Hare-wallaby

mm

A European Rabbit

1   First upper incisors are long, two-columned and curved.
2   Sockets where a very small pair of incisors were present in life.
3   Very long gap between incisors and cheek teeth.
4   There is a network of holes in the bones forming the muzzle.
5   First premolar is flat and single-columned.
6   Third molar is very small. All the rabbit's teeth are open-rooted and grow throughout life. They are often lost after death.

Dental formula: I2, C0, P3, M3
Similar species: Brown Hare

B Northern Brown Bandicoot

7   First four incisors small and crowded, slight space between fourth and fifth. Incisors are often lost after death.
8   Prominent canine.
9   Molars have several sharp cusps, forming triangular patterns.
10 Fourth molar is about three-quarters the size of the other molars.
11 Large, pear-shaped ear bones. (Compare with C) The size and shape of these ear bones is a key diagnostic feature between the two genera of bandicoots, *Isoodon* and *Perameles*.

Dental formula: I5, C1, P3, M4
Similar species: Southern Brown Bandicoot, Golden Bandicoot, Long-nosed Bandicoot (C), Eastern Barred Bandicoot, Rufous Spiny Bandicoot, Bilby

C Long-nosed Bandicoot

12 First four incisors small and crowded, longer space between fourth and fifth incisors than in Northern Brown Bandicoot (B). Incisors often lost after death.
13 Large canine.
14 Molars have several sharp cusps, forming triangular patterns.
15 Fourth molar is about half the size of the other three molars.
16 Ear bones are small and rounded. (Compare with B) Also refer to 11 above.

Dental formula: I5, C1, P3, M4
Similar species: Eastern Barred Bandicoot, Northern Brown Bandicoot (B), Southern Brown Bandicoot, Golden Bandicoot, Rufous Spiny Bandicoot, Bilby

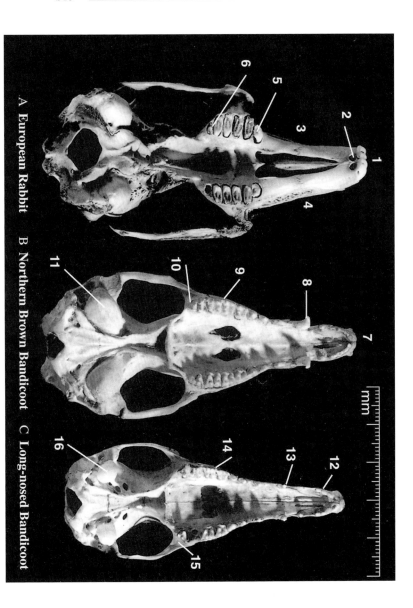

A European Rabbit   B Northern Brown Bandicoot   C Long-nosed Bandicoot

A Common Brushtail Possum

1   First incisors have chisel-shaped ends.
2   Strong canine and first premolar. (This premolar is usually present on both sides of the upper jaw).
3   Palate is ridged.
4   Large blade-like premolar curves outwards.
5   Molars have four sharp cone-shaped cusps.

Dental formula: I3, C1, P2, M4
Similar species: Mountain Brushtail Possum, Northern Brushtail Possum, Spotted Cuscus, Grey Cuscus

B Rufous Bettong

6   First incisors are long and sharp.
7   Second and third incisors are broad; third incisor has a shallow groove about a third of the way along its length.
8   Canine present.
9   Long, broad premolar has several fine grooves.
10 Molars have four low cone-shaped cusps.
11 No holes in rear part of palate. Other bettongs (in the *Bettongia* genus), potoroos and the Musky Rat-kangaroo do have holes at the rear of the palate (C).
12 Skull is short and broad (compare with C).

Dental formula: I3, C1, P1, M4
Similar species: Brush-tailed Bettong, Northern Bettong, Tasmanian Bettong, Long-nosed Potoroo (C), Long-footed Potoroo, Musky Rat-kangaroo

C Long-nosed Potoroo

13 Skull is long and narrow.
14 First upper incisors are long and sharp.
15 Second and third incisors are very small.
16 Canine present.
17 Blade-like premolar is longer than the longest molar and has several fine grooves.
18 Molars have four low cone-shaped cusps.
19 Holes present at the rear of the palate.

Dental formula: I3, C1, P1, M4
Similar species: Long-footed Potoroo, Rufous Bettong (B), Brush-tailed Bettong, Northern Bettong, Tasmanian Bettong, Musky Rat-kangaroo

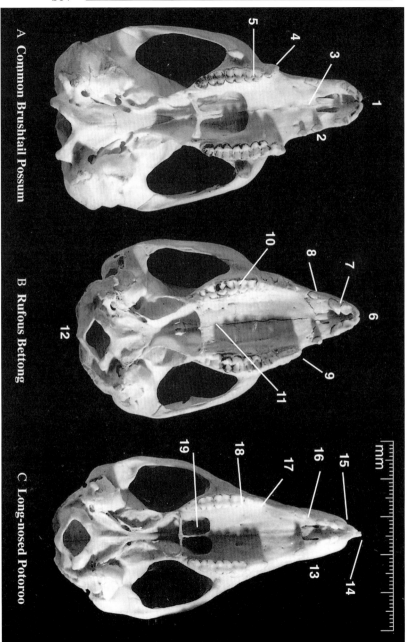

A Common Brushtail Possum

B Rufous Bettong

C Long-nosed Potoroo

A Grey-headed Flying-fox

1 Small incisors, often lost after death.
2 Large, sharp canine.
3 Palate has transverse ridges.
4 Cheek teeth have sharp, pointed cusps.
5 Sharp projection over eye socket.
6 Brain case very large. Seen from the side, this area is curved at an angle to the snout.

Dental formula: I2, C1, P3, M2
Similar species: Little Red Flying-fox, Black Flying-fox, Spectacled Flying-fox

B Greater Glider

7 First incisors are larger than the other two. These are often lost after death.
8 Small canine and first premolar are also often lost.
9 Molars have crescent-shaped ridges which become worn with age.
10 The holes at the rear of the palate are usually smaller than in the ringtail possum (C).
11 This area, where the skull articulates with the lower jaw, is slightly larger than in the very similar ringtail possum skull (C).

Dental formula: I3, C1, P3, M4
Similar species: Common Ringtail Possum (C), Herbert River Ringtail Possum, Green Ringtail Possum, Lemuroid Ringtail Possum, Rock Ringtail Possum, Striped Possum

C Common Ringtail Possum

12 First incisors are larger than the other two. They are often lost after death.
13 Small canine and first premolar are also often lost.
14 Molars have crescent-shaped ridges, which become worn with age.
15 These holes are usually larger than in the Greater Glider (B).
16 This area is slightly smaller than in the Greater Glider (B).

Dental formula: I3, C1, P3, M4
Similar species: Greater Glider (B), Herbert River Ringtail Possum, Green Ringtail Possum, Lemuroid Ringtail Possum, Rock Ringtail Possum, Striped Possum

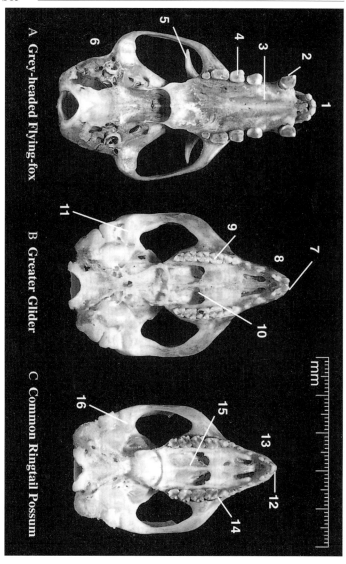

A **Brush-tailed Phascogale**

1 Four pairs of incisors; first pair much larger than the other three pairs.
2 Depression where lower canine fits when mouth is closed.
3 Strong canine.
4 Three sharp premolars.
5 First three molars triangular in shape with sharp cusps.
6 Fourth molar very small.
7 Large rounded ear bones.

Dental formula: I4, C1, P3, M4
Similar species: Red-tailed Phascogale (smaller). Dusky Antechinus (S17) similar but smaller

B **Sugar Glider**

8 First pair of incisors larger than the other two.
9 Small canine present.
10 Molars have four low cone-shaped cusps.
11 No large holes at rear of palate. Small, uneven holes are sometimes present in old skulls.
12 Broad skull.
13 Prominent ear bones.

Dental formula: I3, C1, P3, M4
Similar species: Leadbeater's Possum (slightly smaller). Squirrel Glider, Mahogany Glider, Yellow-bellied Glider all larger

C **Bush Rat** (inset: House Mouse)

14 Single pair of large, curved incisors, with chisel-shaped ends. Hard yellow or orange enamel on front and side surfaces. Incisors continue to grow throughout life.
15 Long gap. No canines or premolars.
16 Three molars, each with several cusps. The size and arrangement of the cusps of the molars varies among the different groups (genera) of rodents, and are two of the features used in the identification of this large and complex family.
17 The small size and notched incisor are diagnostic features of House Mouse.

Dental formula: I1, C0, P0, M3
Similar species: All *Rattus* species and other native rodents generally. The skulls of all rats and mice have many similarities, and most require identification by an expert. The Water-rat and False Water-rat have two molars, not three as in all other rats and mice

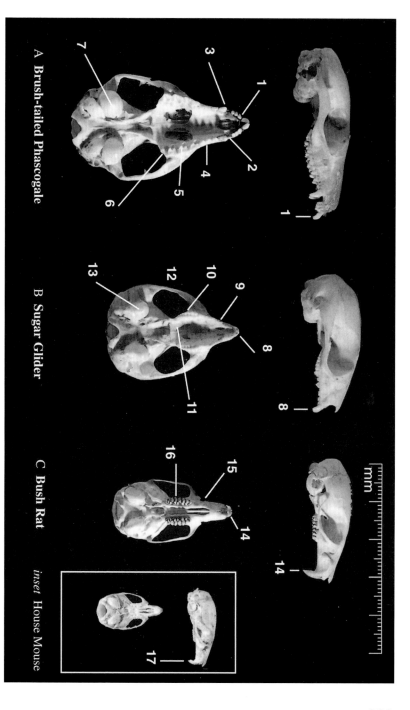

A Brush-tailed Phascogale

B Sugar Glider

C Bush Rat

inset House Mouse

283

Note that the photographs of the skulls shown from underneath are enlarged to twice natural size.

A **Dusky Antechinus**
B **Brown Antechinus**

1 Four pairs of small, crowded incisors, often lost after death.
2 Canine present.
3 Three pairs of crowded premolars.
4 First three molars triangular in shape with sharp cusps, fourth molar small.
5 These holes in the front of the palate are larger in Dusky Antechinus (A) than in Brown Antechinus (B).

Dental formula: Dusky Antechinus — I4, C1, P3, M4
Brown Antechinus — I4, C1, P3, M4
Similar species: Other *Antechinus* species; Red-tailed Phascogale; Brush-tailed Phascogale similar but larger (S16). Also dunnarts, planigales and other small insectivorous marsupials

C **Eastern Pygmy-possum**

6 First incisors point slightly forward, others very small.
7 Canine present.
8 Large third premolar, others very small, peg-like.
9 Four low, cone-shaped cusps on the three molars. (Most marsupials have four molars). The second and third molars have been lost in this specimen.
10 Broad skull.
11 Prominent ear bones.

Dental formula: I3, C1, P3, M3
Similar species: Western Pygmy-possum, Little Pygmy-possum, Long-tailed Pygmy-possum, Mountain Pygmy-possum (has greatly enlarged premolar), Feathertail Glider

D **Gould's Wattled Bat**

12 Deep rounded hollow.
13 Short, broad snout; incisors very small.
14 Large canine.
15 Sharp cusps on all cheek teeth.
16 Third molar small.

Dental formula: I2, C1, P1, M3
Similar species: Chocolate Wattled Bat and all other *Chalinolobus* species. Skulls of all small insectivorous bats have many similarities and require identification by an expert

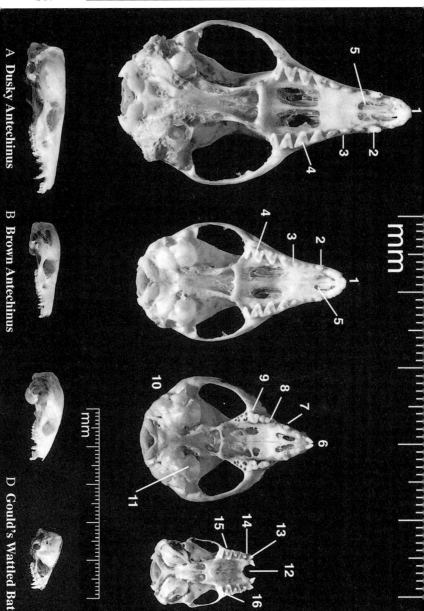

A Dusky Antechinus

B Brown Antechinus

D Gould's Wattled Bat

mm

A **Pig**

1 Strong, forward-pointing incisors — second and third pairs have been lost in this specimen.
2 Canine missing in this young specimen — large in adult animal.
3 Not all the cheek teeth have erupted. Molars have many knobbed cusps.
4 Short projection.
5 This area is almost flat, as the pig is a placental mammal. In marsupials this part of the jaw bends inwards to form a prominent shelf.

Dental formula: I2-3, C1, P4, M3
Similar species: None

B **Goat**

6 Incisors often lost after death, as in this specimen. The canine, also lost, is a similar shape to the incisors and lies beside the third incisor.
7 Long gap.
8 Cheek teeth have crescent-shaped ridges.
9 Long projection, more acute angle to jaw than in sheep (C).
10 No shelf, as the Goat is a placental mammal.

Dental formula: I3, C1, P3, M3
Similar species: Sheep, some species of deer

C **Sheep**

11 Three pairs of incisors, often lost after death. Canine lies beside incisors, also often lost.
12 Long gap.
13 Cheek teeth have sharp crescent-shaped ridges.
14 Long projection, at a larger angle to the jaw than in Goat (B).
15 No shelf, as the sheep is a placental mammal.

Dental formula: I3, C1, P3, M3
Similar species: Goat, some species of deer

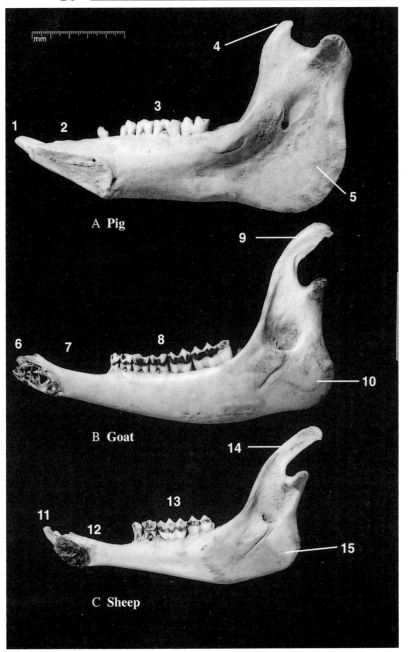

A Pig

B Goat

C Sheep

Eastern Grey Kangaroo

1   Large blade-like, forward-pointing incisor in each jaw.
2   Rough area where jaws were joined in life.
3   Small blade-like premolar.
4   Lower molars move forward in the jaw (see skull, S3).
5   A hole through the bone in this area is typical of all members of the kangaroo family.
6   Prominent shelf. The inflexion of this area of the jaw is characteristic of marsupials.

Dental formula: I1, C0, P1, M4
Similar species: Western Grey Kangaroo, Red Kangaroo, Common Wallaroo, Antilopine Wallaroo, Black Wallaroo. Also Large Wallabies — see Swamp Wallaby and Red-necked Wallaby (D5).

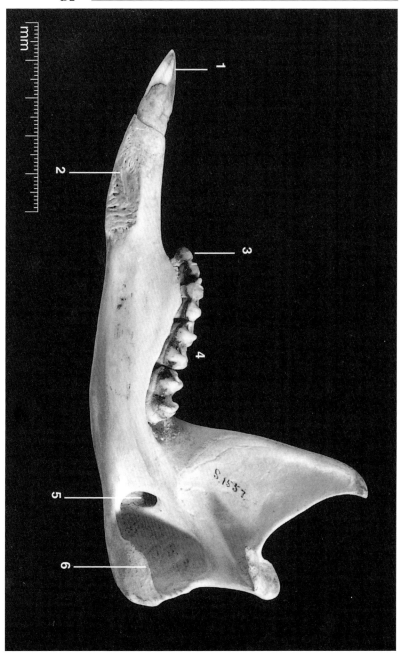

A **Red Fox**

1   Three small incisors.
2   Large strong canine.
3   First molar is the largest cheek tooth.
4   No shelf, as the fox is a placental mammal

Dental formula: I3, C1, P4, M3
Similar species: Dog/Dingo (almost identical to fox), Tasmanian Devil (C), Thylacine. Cat (smaller) has only 3 cheek teeth.

This is the lower jaw of a female; the jaw is much larger in the male seal and the canines are larger.

B **Australian Fur-seal**

5   Two strong incisors.
6   Large strong canine.
7   All cheek teeth have pointed cusps.
8   No shelf, as the seal is a placental mammal.

Dental formula: I2, C1, P&M5

C **Tasmanian Devil**

9   Three crowded incisors.
10   Large strong canine.
11   All four molars about the same size, but large, with strong sharp cusps.
12   Prominent shelf.

Dental formula: I3, C1, P2, M4
Similar species: Dog/ Dingo, Red Fox (A), Thylacine

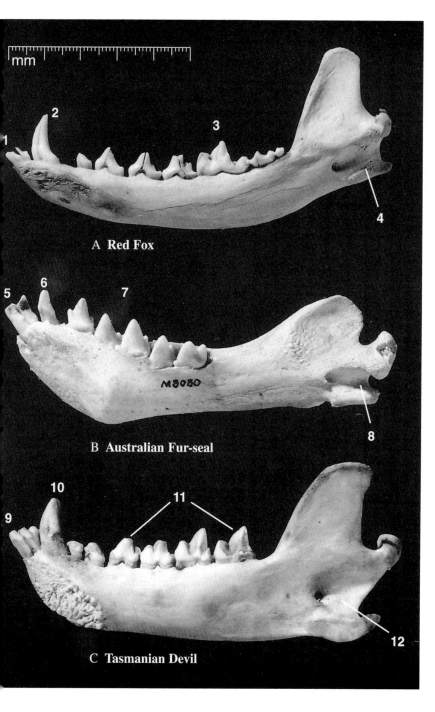

A **Red Fox**

B **Australian Fur-seal**

C **Tasmanian Devil**

291

Note that the photographs have been reduced to 2/3 natural size.

Both these species have strongly fused lower jaws which are usually found joined.

A **Common Wombat**

1 Single pair of forward-pointing incisors, with chisel-shaped ends.
2 Cheek teeth have flattened crowns; sharp ridge of enamel surrounds softer dentine. Teeth often lost after death.
3 Prominent shelf.

Dental formula: I1, C0, P1, M4
Similar species: Southern Hairy-nosed Wombat, Northern Hairy-nosed Wombat

B **Koala**

4 Single pair of forward-pointing incisors, smaller than in wombats. Chisel-shaped ends.
5 Cheek teeth have sharp crescentic ridges which become worn with age.
6 Shelf present but unusually small for a marsupial.

Dental formula: I1, C0, P1, M4
Similar species: none

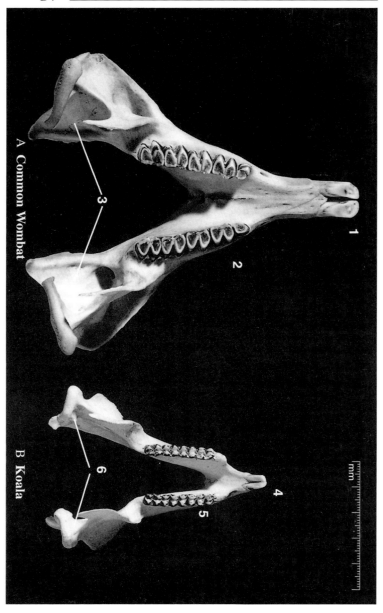

293

Lower jaws of large male wallabies may be similar in size to jaws of female kangaroos.

A Swamp Wallaby

1   Single blade-like, forward-pointing incisor in each jaw.
2   Blade-like premolar.
3   Angle of jaw almost 90°.
4   This projection wider and more robust than in other large wallabies.
5   Hole through the bone, characteristic of kangaroo family.
6   Prominent shelf.

Dental formula: I1, C0, P1, M4
Similar species: Red-necked Wallaby (B), Agile Wallaby, Black-striped Wallaby, Whiptail Wallaby, Western Brush Wallaby, Tammar Wallaby. Parma Wallaby similar but smaller.

B Red-necked Wallaby

7   Single blade-like, forward-pointing incisor in each jaw.
8   Premolar is small and shorter than the first molar.
9   Angle of the jaw greater than in Swamp Wallaby.
10   Hole through the bone.
11   Prominent shelf, but not as prominent as Swamp Wallaby.

Dental formula: I1, C0, P1, M4
Similar species: Agile Wallaby, Black-striped Wallaby, Whiptail Wallaby, Western Brush Wallaby, Tammar Wallaby, Swamp Wallaby (A). Parma Wallaby similar but smaller.

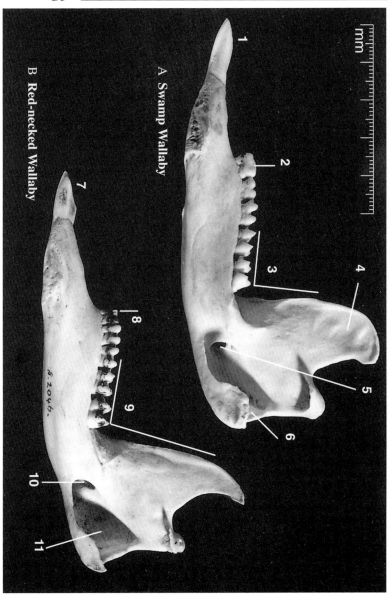

A Swamp Wallaby

B Red-necked Wallaby

Lower jaws of young large wallabies are similar in size (see young Agile Wallaby, D7).

A **Red-necked Pademelon**

1 Single blade-like, forward-pointing incisor in each jaw.
2 Blade-like premolar.
3 Angle of jaw about 90°.
4 Hole through the bone.
5 Prominent shelf.

Dental formula: I1, C0, P1, M4
Similar species: Red-legged Pademelon, Tasmanian Pademelon, Brush-tailed Rock-wallaby (B) and other rock-wallabies, Northern Nailtail Wallaby (C), Bridled Nailtail Wallaby, Lumholtz's Tree-kangaroo (D7), Bennett's Tree-kangaroo, Spectacled Hare-wallaby (D7), Quokka.

B **Brush-tailed Rock-wallaby**

6 Single blade-like, forward-pointing incisor in each jaw.
7 Blade-like premolar.
8 Angle of jaw about 90°.
9 Hole through the bone.
10 Prominent shelf.

Dental formula: I1, C0, P1, M4
Similar species: Other species of rock-wallaby, Red-necked Pademelon (A), Red-legged Pademelon, Tasmanian Pademelon, Northern Nailtail Wallaby (C), Bridled Nailtail Wallaby, Lumholtz's Tree-kangaroo (D7), Bennett's Tree-kangaroo, Spectacled Hare-wallaby (D7), Quokka

C **Northern Nailtail Wallaby**

11 Single blade-like, forward-pointing incisor in each jaw.
12 Very small premolar.
13 Angle of jaw greater than most other small wallabies.
14 Hole through the bone.
15 Prominent shelf.

Dental formula: I1, C0, P1, M4
Similar species: Bridled Nailtail Wallaby, Red-necked Pademelon (A), Red-legged Pademelon, Tasmanian Pademelon, Brush-tailed Rock-wallaby (B) and other rock-wallabies, Lumholtz's Tree-kangaroo (D7), Bennett's Tree-kangaroo, Spectacled Hare-wallaby (D7), Quokka

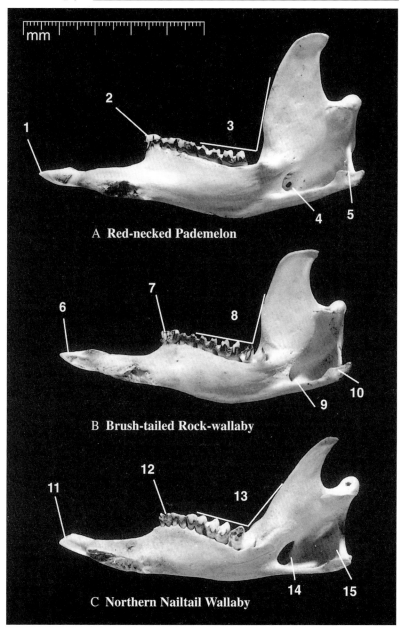

mm

A Red-necked Pademelon

B Brush-tailed Rock-wallaby

C Northern Nailtail Wallaby

A Spectacled Hare-wallaby

1   Single blade-like, forward-pointing incisor in each jaw.
2   Premolar about the same size as first molar.
3   Angle of jaw about 90°.
4   Hole through the bone.
5   Prominent shelf.

Dental formula: I1, C0, P1, M4
Similar species: Red-necked Pademelon (D6), Red-legged Pademelon, Tasmanian Pademelon, Brush-tailed Rock-wallaby (D6) and other rock-wallabies, Northern Nailtail Wallaby (D6), Bridled Nailtail Wallaby, Lumholtz's Tree-kangaroo (B), Bennett's Tree-kangaroo, Quokka. Lower jaws of young wallabies are similar in size (see young Agile Wallaby, C).

B Lumholtz's Tree-kangaroo

6   Single blade-like, forward-pointing incisor in each jaw.
7   Premolar a long, ribbed, cutting tooth.
8   Fourth molar not fully erupted.
9   Angle of jaw about 100°.
10   Hole through the bone.
11   Prominent shelf.

Dental formula: I1, C0, P1, M4
Similar species: Bennett's Tree-kangaroo, Red-necked Pademelon (D6), Red-legged Pademelon, Tasmanian Pademelon, Brush-tailed Rock-wallaby (D6) and other rock-wallabies, Northern Nailtail Wallaby (D6), Bridled Nailtail Wallaby, Spectacled Hare-wallaby (A), Quokka. Lower jaws of young wallabies are similar in size (see young Agile Wallaby, C).

C Agile Wallaby (young)

Compare this jaw with adult jaw of Red-necked Wallaby (D5).

Lower jaw of a young large wallaby is included here as it is comparable in size with the lower jaws of small wallabies and can easily be mistaken for them. Note that not all the cheek teeth have erupted (12). This is the most useful diagnostic feature of young macropods — the molar teeth that have not yet erupted can be seen in the jaw below the gum line.

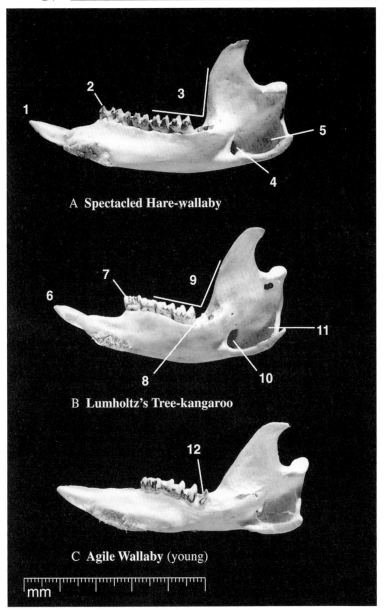

A **Spectacled Hare-wallaby**

B **Lumholtz's Tree-kangaroo**

C **Agile Wallaby** (young)

A Spot-tailed Quoll

1 Three small incisors.
2 Large canine.
3 Sharp cusps on cheek teeth.
4 Prominent shelf (compare with Red Fox, D3).

Dental formula: I3, C1, P2, M4
Similar species: Eastern Quoll, Northern Quoll, Western Quoll, Tasmanian Devil (D3)

B Common Brushtail Possum

5 Single large incisor in each jaw, with chisel-shaped end.
6 Very small second incisor may be present.
7 Large premolar curves outwards.
8 Molars have four sharp cone-shaped cusps.
9 Angle made with the jaw is greater than in ringtail possums (D10).
10 Prominent shelf.

Dental formula: I2, C0, P1, M4
Similar species: Mountain Brushtail Possum, Northern Brushtail Possum, Spotted Cuscus, Grey Cuscus

C Rufous Bettong

11 Single blade-like, forward-pointing incisor in each jaw.
12 Long blade-like premolar has several fine grooves.
13 Molars have four cone-shaped cusps. Fourth molar has not yet erupted in this specimen.
14 Angle of jaw is smaller than in potoroo (D9).
15 Prominent shelf.

Dental formula: I1, C0, P1, M4
Similar species: Brush-tailed Bettong, Northern Bettong, Tasmanian Bettong, Long-nosed Potoroo (D9), Long-footed Potoroo. The Musky Rat-kangaroo has a very small second incisor (often lost after death), and a very distinctive premolar.

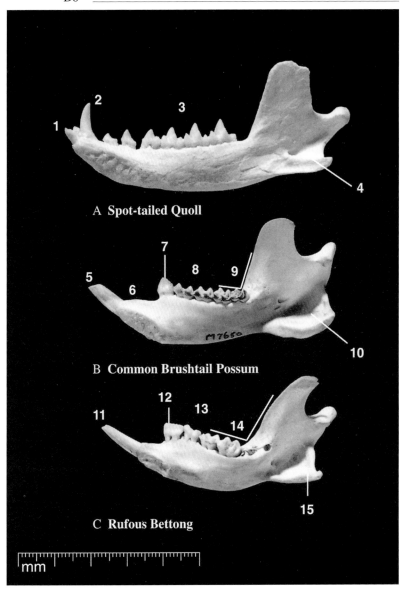

A Spot-tailed Quoll

B Common Brushtail Possum

C Rufous Bettong

A **European Rabbit**

1   Single strong incisor, curved, with chisel-shaped end.
2   Long gap.
3   Cheek teeth are double-columned, with sharp cusps. These are often lost after death.
4   Shape of projection is characteristic of rabbit and hare.
5   No shelf — rabbit is a placental mammal.
Dental formula: I1, C0, P2, M3
Similar species: Brown Hare

B **Long-nosed Bandicoot**

6   Three pairs of small incisors in each jaw, project forward. These are often lost after death.
7   Canine present.
8   Cheek teeth have sharp cusps.
9   Angle of jaw greater than in Northern Brown Bandicoot (C).
10   Shelf has a sharp projection.
Dental formula: I3, C1, P3, M4
Similar species: Eastern Barred Bandicoot, Northern Brown Bandicoot (C), Southern Brown Bandicoot, Golden Bandicoot, Rufous Spiny Bandicoot, Bilby

C **Northern Brown Bandicoot**

11   Three pairs of small incisors in each jaw, project forward. These are often lost after death.
12   Canine present. (Usually sharp – tip broken in photograph.)
13   Cheek teeth have sharp cusps.
14   Angle of jaw less than Long-nosed Bandicoot (B).
15   Shelf has a sharp projection.
Dental formula: I3, C1, P3, M4
Similar species: Southern Brown Bandicoot, Golden Bandicoot, Long-nosed Bandicoot (B), Eastern Barred Bandicoot, Rufous Spiny Bandicoot, Bilby

D **Long-nosed Potoroo**

16   Single blade-like, forward-pointing incisor in each jaw.
17   Long blade-like premolar with several fine grooves.
18   Molars have four low, cone-shaped cusps.
19   Angle of jaw larger than in Rufous Bettong (D8).
20   Prominent shelf.
Dental formula: I1, C0, P1, M4
Similar species: Long-footed Potoroo, Rufous Bettong (D8), Brush-tailed Bettong, Northern Bettong, Tasmanian Bettong. Musky Rat-kangaroo has small second incisor and distinctive premolar.

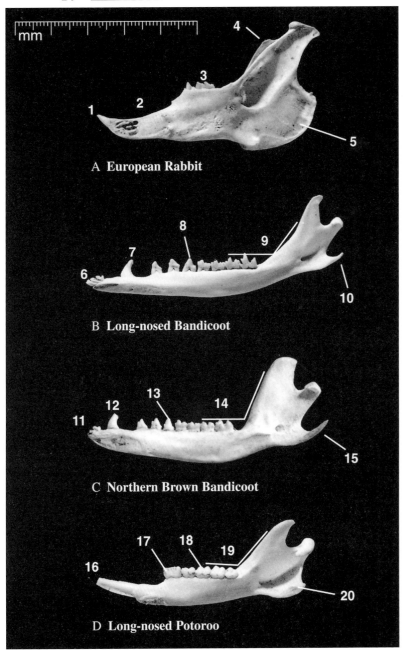

A European Rabbit

B Long-nosed Bandicoot

C Northern Brown Bandicoot

D Long-nosed Potoroo

A Common Ringtail Possum

1   Blade-like, forward pointing incisor (compare with brushtail possum, D8).
2   Small socket where one, or sometimes two or three, very small incisors may be present — often lost after death.
3   Cheek teeth have small, sharp, crescent-shaped ridges.
4   Angle made with the jaw nearly 90°.
5   Prominent shelf with rounded projection.

Dental formula: I2-4, C0, P1, M4
Similar species: Greater Glider (B), Herbert River Ringtail Possum, Green Ringtail Possum, Lemuroid Ringtail Possum, Rock Ringtail Possum, Striped Possum

B Greater Glider

6   Blade-like, forward pointing incisor.
7   Socket where very small incisor was present.
8   Cheek teeth have small crescent-shaped ridges.
9   Angle made with jaw slightly greater than ringtail (A).
10 Shelf has a more pointed projection than ringtail (A).

Dental formula: I2, C0, P1, M4
Similar species: Common Ringtail Possum (A), Herbert River Ringtail Possum, Green Ringtail Possum, Lemuroid Ringtail Possum, Rock Ringtail Possum, Striped Possum

C Grey-headed Flying-fox

11 Two small incisors, often lost after death.
12 Large canine.
13 Cheek teeth have sharp, pointed cusps.
14 Large angle.
15 No shelf — flying-fox is a placental mammal.

Dental formula: I2, C2, P3, M3
Similar species: Little Red Flying-fox, Black Flying-fox, Spectacled Flying-fox

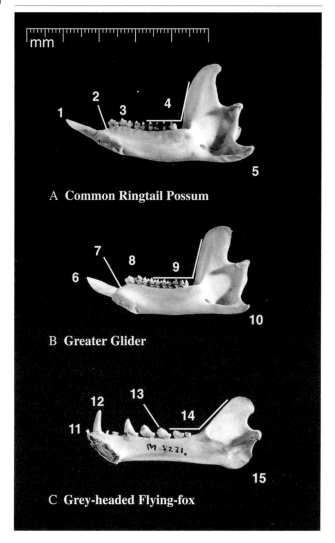

A Common Ringtail Possum

B Greater Glider

C Grey-headed Flying-fox

A **Brush-tailed Phascogale**

1 First incisor larger than other two.
2 Large canine.
3 Sharp cusps on molars.
4 Prominent shelf with pointed projection.

Dental formula: I3, C1, P2, M4
Similar species: Red-tailed Phascogale (smaller). Dusky Antechinus (D12) similar but much smaller.

B **Bush Rat**

5 Single large, curved incisor in each jaw. Hard yellow or orange enamel on front and side surfaces. Incisors continue to grow throughout life and are often lost after death.
6 Three molars, each with several cusps.
7 Small sharp projection.
8 No shelf — rats are placental mammals. This projection is in the same plane as the rest of the jaw.

Dental formula: I1, C0, P 0, M3
Similar species: All *Rattus* species and other native rodents generally. The lower jaws of all rats and mice have many similarities, and most require identification by an expert. The Water-rat and False Water-rat have two molars, not three as in all other rats and mice.

C **Sugar Glider**

9 Single blade-like, forward pointing incisor in each jaw.
10 Four small premolars present in life, but often lost after death. Only one present in this specimen.
11 Molars have four low cone-shaped cusps.
12 Sharp, pointed projection.
13 Prominent shelf with sharp projection.

Dental formula: I1, C0, P4, M4
Similar species: Leadbeater's Possum (slightly smaller). Squirrel Glider, Mahogany Glider, Yellow-bellied Glider (all larger).

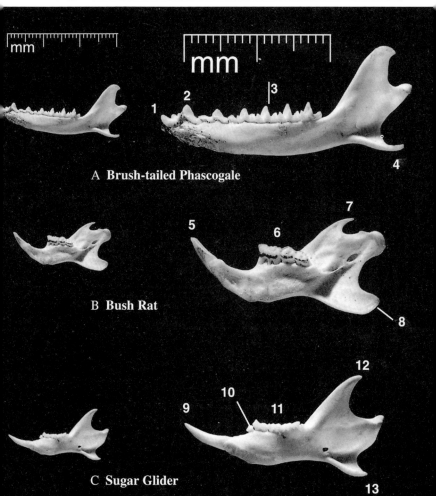

A **Brush-tailed Phascogale**

B **Bush Rat**

C **Sugar Glider**

Note that each of these lower jaws is shown natural size and twice natural size.

A Dusky Antechinus

1   Three small incisors in each jaw, often lost after death, as in this specimen.
2   Canine present.
3   Cheek teeth have sharp cusps.
4   Prominent shelf with sharp projection.

Dental formula: I3, C1, P3, M4
Similar species: Brown Antechinus (B) and other *Antechinus* species; Red-tailed Phascogale; Brush-tailed Phascogale similar but larger.

B Brown Antechinus

5   Three small incisors in each jaw, often lost after death.
6   Small canine.
7   Cheek teeth crowded, with sharp cusps.
8   Prominent shelf with sharp projection.

Dental formula: I3, C1, P3, M4
Similar species: Dusky Antechinus (A) and other *Antechinus* species; also dunnarts, planigales and other small insectivorous marsupials.

C Eastern Pygmy-possum

9   Single forward-pointing incisor in each jaw.
10   Three small premolars, often lost after death, and one larger premolar.
11   Three molars have four low cone-shaped cusps.
12   Prominent shelf with a sharp projection.

Dental formula: I1, C0, P4, M3
Similar species: Western Pygmy-possum, Little Pygmy-possum (has four molars), Long-tailed Pygmy-possum, Mountain Pygmy-possum (has greatly enlarged premolar), Feathertail Glider.

D Gould's Wattled Bat

13   Three very small incisors, often lost after death.
14   Strong canine present.
15   Cheek teeth have sharp cusps.
16   No shelf — bats are placental mammals.

Dental formula: I3, C1, P2, M3
Similar species: Chocolate Wattled Bat and all other *Chalinolobus* species. Lower jaws of all small bats are very similar.

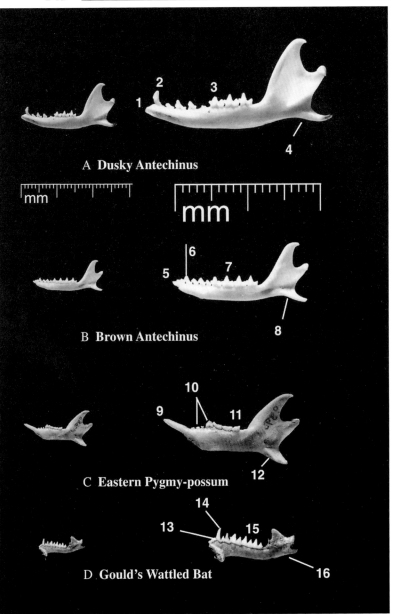

A **Dusky Antechinus**

B **Brown Antechinus**

C **Eastern Pygmy-possum**

D **Gould's Wattled Bat**

A **Dog/Dingo**

This humerus is from a Dingo. There is much variation in size in domestic and feral Dogs.

1 Nearly flat front, with narrow ridge.
2 Long shaft.
3 Hole near centre of base.
4 Relatively narrow base.

Similar species: Red Fox and Cat (smaller), Thylacine

B **Eastern Grey Kangaroo**

This is from a female; humerus of male kangaroo is about 18 cm long.

5 Long ridge with prominent crest.
6 Prominent muscle attachment area.
7 Flange ends in a point.
8 Hole behind bar of bone. This hole is found in most marsupials.

Similar species: Western Grey Kangaroo, Red Kangaroo, Common Wallaroo, Antilopine Wallaroo, Black Wallaroo. Also, the humerus of a large male wallaby is about the same size as that of a small female kangaroo.

C **Swamp Wallaby**

This is from a female; humerus of male wallaby is about 11 cm long.

9 Long ridge with prominent crest.
10 Prominent muscle attachment area.
11 Flange ends in a point.
12 Hole behind bar of bone.

Similar species: Red-necked Wallaby, Agile Wallaby, Black-striped Wallaby, Whiptail Wallaby, Western Brush Wallaby, Tammar Wallaby. Parma Wallaby similar but smaller. The humerus of a small female kangaroo is about the same size as that of a large male wallaby.

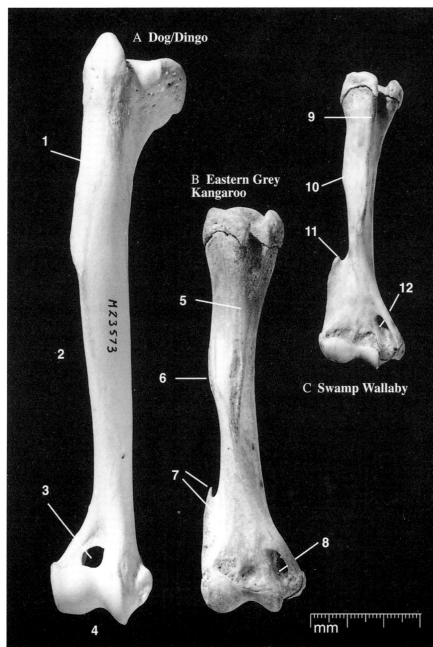

A Dog/Dingo

B Eastern Grey Kangaroo

C Swamp Wallaby

311

A **Common Wombat**

1 Very robust head.
2 Large prominence where muscle is attached in life.
3 Large flange with pointed end.
4 Depression.
5 Hole present behind bar of bone.
6 Very wide base.

Similar species: Southern Hairy-nosed Wombat, Northern Hairy-nosed Wombat

B **Koala**

7 Prominent ridge.
8 Roughened area where muscle is attached in life.
9 Flange with pointed end.
10 Hole present behind bar of bone.
11 Shallow depression.

Similar species: Some similarities to kangaroos and wallabies, but ridge (7) is more prominent in Koala

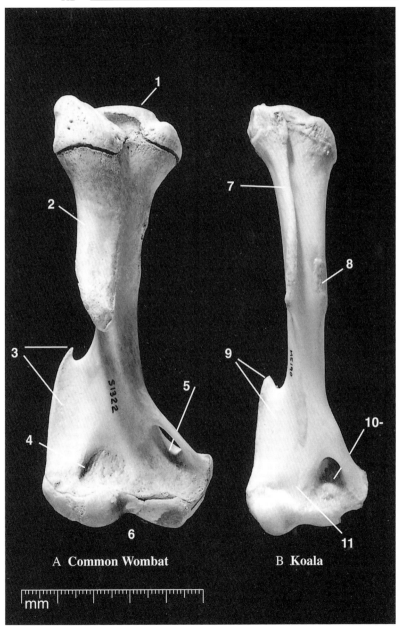

A Common Wombat

B Koala

mm

A **Greater Glider**

1 Ridge short with sharp crest.
2 Long slender shaft.
3 Narrow flange with sharp point.
4 Hole under bar of bone.
5 Narrow base.

Similar species: Some similarities to all possums and gliders, but the slender shaft (2) and narrow base (5) are characteristic of Greater Glider

B **Spot-tailed Quoll**

6 Flat front.
7 No point on flange.
8 No hole (unusual in marsupials).

Similar species: Eastern Quoll, Northern Quoll, Western Quoll

C **Common Brushtail Possum**

9 Very prominent ridge with sharp crest.
10 Wide flange with sharp point.
11 Hole under bar of bone.

Similar species: Mountain Brushtail Possum, Northern Brushtail Possum, Spotted Cuscus, Grey Cuscus. Humerus of young brushtail very similar to humerus of Common Ringtail Possum (H4).

D **European Rabbit**

12 Small flat front.
13 Short ridge.
14 Hole near centre of base, but not under bar of bone.
15 Narrow base with prominent grooves.

Similar species: Brown Hare

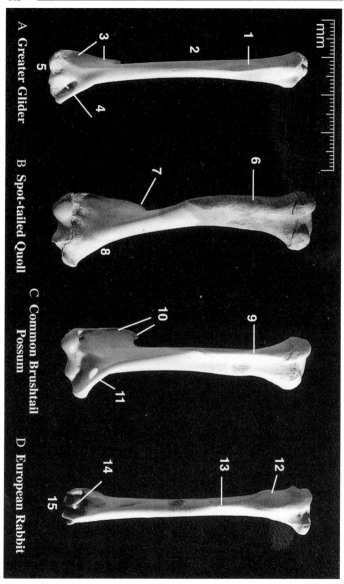

A Greater Glider B Spot-tailed Quoll C Common Brushtail Possum D European Rabbit

A **Long-nosed Bandicoot**

1   Prominent flat front.
2   Robust shaft.
3   Hole in centre of base.
4   Hole under bar of bone.

Similar species: Eastern Barred Bandicoot, Northern Brown Bandicoot, Southern Brown Bandicoot, Golden Bandicoot, Rufous Spiny Bandicoot, Bilby

B **Common Ringtail Possum**

5   Prominent ridge with sharp crest.
6   Wide flange with sharp point.
7   Hole under bar of bone.

Similar species: Herbert River Ringtail Possum, Green Ringtail Possum, Lemuroid Ringtail Possum, Rock Ringtail Possum. Also similar to humerus of young brushtail possum (H3). Yellow-bellied Glider similar size, but ridge (5) shorter and flange (6) narrower than in ringtail.

C **Rufous Bettong**

8   Very prominent flat front.
9   Wide flange with point.
10   No hole under bar of bone — a very unusual feature. This hole is found in other bettongs (*Bettongia*) but not in the Rufous Bettong (*Aepyprymnus*).

Similar species: Brush-tailed Bettong, Tasmanian Bettong, Long-nosed Potoroo (D), Long-footed Potoroo, Musky Rat-kangaroo

D **Long-nosed Potoroo**

11   Very prominent flat front.
12   Long flange with point (flange narrower than in bettongs (C)).
13   Hole behind bar of bone.

Similar species: Long-footed Potoroo, Rufous Bettong (C), Brush-tailed Bettong, Tasmanian Bettong, Musky Rat-kangaroo

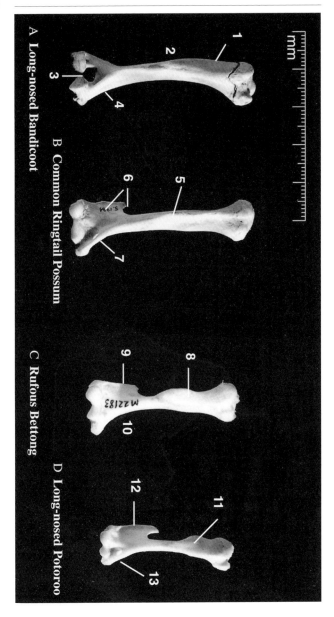

A Long-nosed Bandicoot B Common Ringtail Possum C Rufous Bettong D Long-nosed Potoroo

Note that each of these bones is shown natural size (A1, B1, C1, D1, E1) and twice natural size (A, B, C, D, E)

A Sugar Glider

1   Narrow ridge.
2   Long slender shaft.
3   Long flange with sharp point.
4   Hole under bar of bone.

Similar species: Leadbeater's Possum (slightly smaller). Squirrel Glider, Mahogany Glider, Yellow-bellied Glider, Striped Possum (all larger)

B Bush Rat

5   Ridge with very prominent crest.
6   Deep hollow, sometimes with hole in centre.
7   Small hole under thin bar of bone.
8   Wide base.

Similar species: All *Rattus* species. The humeri of all rats and mice have many similarities.

C Dusky Antechinus

9   Humerus large, compared with Brown Antechinus (D).
10   Prominent ridge with sharp crest, appears twisted to one side.
11   Well-developed flange with point.
12   Hole under bar of bone.

Similar species: Other *Antechinus* species

D Brown Antechinus

13   Inconspicuous ridge.
14   Slender shaft.
15   Narrow flange with no definite point — compare with C and E.
16   Deep hollow, sometimes with hole in centre.
17   Hole under bar of bone.

Similar species: Other *Antechinus* species. Also dunnarts, planigales and other small insectivorous marsupials.

E Eastern Pygmy-possum

18   Sharp ridge.
19   Well-developed flange with point.
10   Hole under bar of bone.

Similar species: Western Pygmy-possum, Little Pygmy-possum, Long-tailed Pygmy-possum, Mountain Pygmy-possum, Feathertail Glider

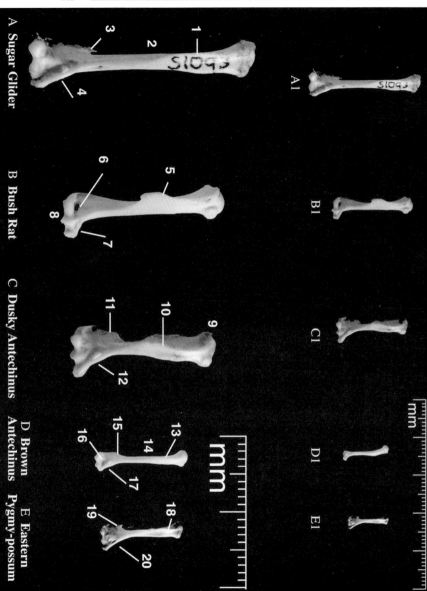

A Sugar Glider
1
2
3
4

A1

B Bush Rat
5
6
7
8

B1

C Dusky Antechinus
9
10
11
12

C1

D Brown Antechinus
13
14
15
16
17

D1

E Eastern Pygmy-possum
18
19
20

E1

mm

319

A Swamp Wallaby

1 Projection is considerably higher than articulating knob. This is a characteristic of all kangaroos and wallabies.
2 Knob fairly flat across the top.
3 Long groove.
4 Rounded edge to flange.
5 Rough projection where muscle is attached in life.
6 Depression on side, near bottom of shaft.

Similar species: Red-necked Wallaby, Agile Wallaby, Black-striped Wallaby, Whiptail Wallaby, Western Brush Wallaby, Tammar Wallaby. Parma Wallaby smaller. Femurs of kangaroos also similar but larger — femur of male Eastern Grey Kangaroo about 27-30 cm long.

B Dog/Dingo

This femur is from a Dingo. There is much variation in size in domestic and feral Dogs.

7 Projection slightly lower than the knob.
8 Large round knob.
9 Wide groove.
10 Well-developed projection.
11 Wide groove at base.
12 Two slight ridges running down shaft.
13 Rough muscle scars.

Similar species: Red Fox and Cat (smaller), Thylacine.

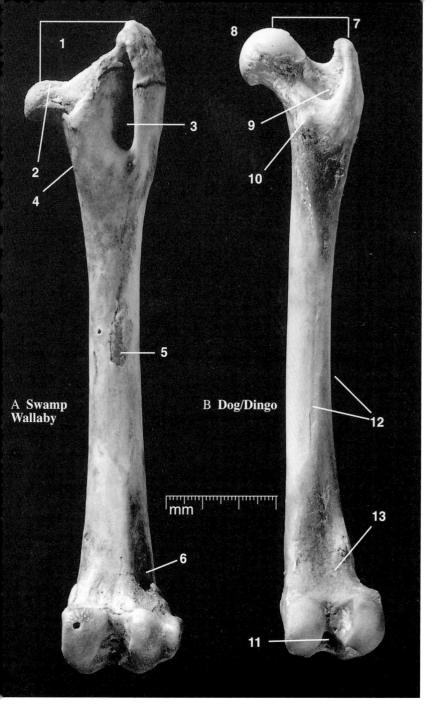

A **Swamp Wallaby**

B **Dog/Dingo**

321

A **Common Wombat**

1 Projection and knob at about the same level.
2 Large round knob.
3 Projection with smooth top.
4 Long deep groove.
5 Long ridge.
6 Muscle scar.
7 Shallow depression.
8 Wide base with deep groove.

Similar species: Southern Hairy-nosed Wombat, Northern Hairy-nosed Wombat.

B **Koala**

9 Projection slightly higher than knob.
10 Large round knob.
11 Projection with smooth top.
12 Deep groove.
13 Muscle scar.

Similar species: Some similarities to Dog (F1), but knob (10) lower and groove (12) narrower than in Dog.

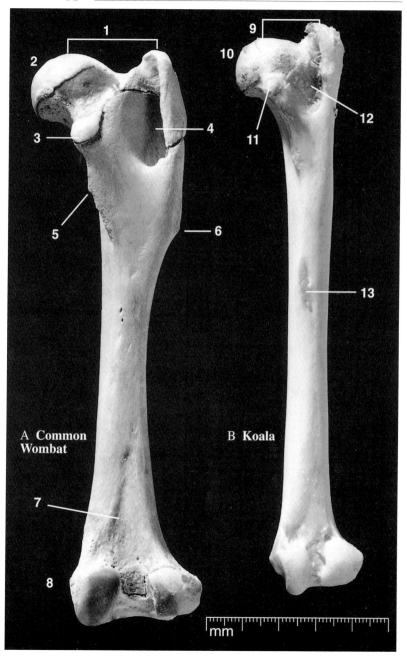

A Common Wombat

B Koala

mm

A Rufous Bettong

1 Projection higher than knob.
2 Knob fairly flat across the top.
3 Deep groove.
4 Rounded edge to flange.
5 Muscle scar.
6 Shaft wide near base.

Similar species: Brush-tailed Bettong, Tasmanian Bettong, Long-footed Potoroo, Long-nosed Potoroo (F4) smaller.

B Greater Glider

7 Projection and knob at about the same level (top of projection has been worn away in this weathered specimen).
8 Round knob.
9 Short, deep groove.
10 Prominent flange.
11 Small muscle scar.
12 Long slender shaft.
13 Narrow base.

Similar species: Some similarities to all possums and gliders, but the slender shaft (12) and narrow base (13) are characteristic of Greater Glider.

C Spot-tailed Quoll

14 Projection slightly higher than knob.
15 Round knob.
16 Deep groove.
17 Prominent flange.
18 Shaft wide near base

Similar species: Eastern Quoll, Northern Quoll, Western Quoll. Also similar to brushtail possums (F4) but projection (14) lower and flange (17) less prominent in quolls.

D European Rabbit

19 Sloping projection, higher than knob.
20 Knob relatively small.
21 Deep groove.
22 Two projections.
23 Slight ridge runs down both sides of shaft.
24 Lower end of shaft has a distinct curve when viewed from the side.

Similar species: Brown Hare.

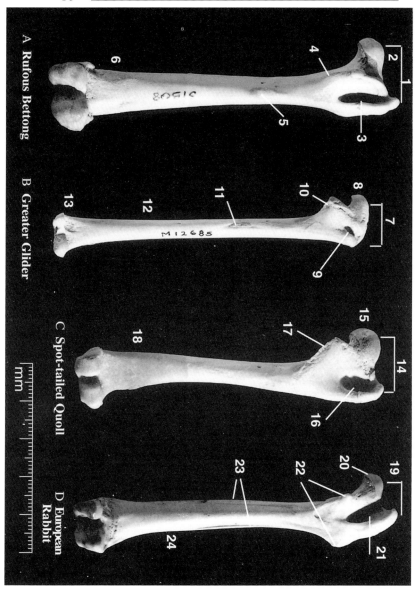

A Rufous Bettong

B Greater Glider

C Spot-tailed Quoll

D European Rabbit

A Common Brushtail Possum

1   Projection higher than knob.
2   Round knob.
3   Prominent flange with rounded top.
4   Deep groove.
5   Muscle scar.

Similar species: Mountain Brushtail Possum, Northern Brushtail Possum, Spotted Cuscus, Grey Cuscus. Femur of young brushtail very similar to femur of Common Ringtail Possum (C) but flange (3) less prominent in brushtail.

B Long-nosed Bandicoot

6   Rounded projection slightly higher than knob.
7   Round knob.
8   Very short flange with large projection.
9   Deep groove.
10 Small round muscle attachment.
11 Viewed from other side, or 'front' of femur, there is a long groove at the base, characteristic of bandicoots.

Similar species: Eastern Barred Bandicoot, Northern Brown Bandicoot, Southern Brown Bandicoot, Golden Bandicoot, Rufous Spiny Bandicoot, Bilby. Also similar to potoroo (D).

C Common Ringtail Possum

12 Projection slightly higher than knob.
13 Round knob.
14 Very prominent pointed flange.
15 Deep groove.

Similar species: Herbert River Ringtail Possum, Green Ringtail Possum, Lemuroid Ringtail Possum, Rock Ringtail Possum. Also similar to femur of young brushtail possum (A).

D Long-nosed Potoroo

16 Projection much higher than knob.
17 Knob fairly flat across the top.
18 Long deep groove.
19 Small flange with sharp edge.
20 Projection.
21 Small, round, muscle scar on shaft (similar to bandicoot, B).

Similar species: Long-footed Potoroo, Rufous Bettong (F3), Brush-tailed Bettong, Tasmanian Bettong, Musky Rat-kangaroo. Also similar to bandicoot (B) but flange (19) is narrower and the groove at 'front' of base of femur not as long as in bandicoot.

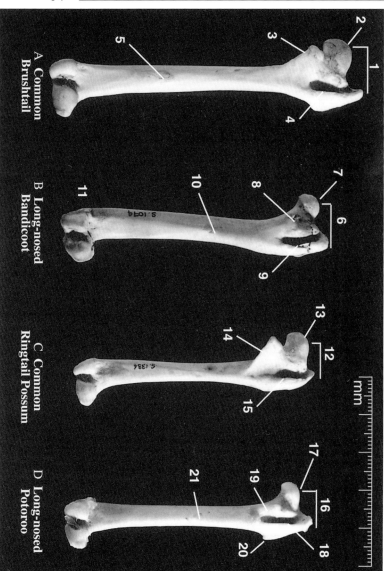

A Common Brushtail

B Long-nosed Bandicoot

C Common Ringtail Possum

D Long-nosed Potoroo

327

A **Sugar Glider**

1 Projection and knob at about the same level.
2 Groove not obvious in this weathered specimen.
3 Prominent flange.
4 Long slender shaft.

Similar species: Leadbeater's Possum (smaller). Squirrel Glider, Mahogany Glider, Yellow-bellied Glider, Striped Possum (all larger).

B **Bush Rat**

5 Projection higher than knob.
6 Very prominent knob on narrow neck.
7 Deep groove.
8 Well-developed crest, gives femur a tilted appearance.

Similar species: All *Rattus* species. The femurs of all rats and mice have many similarities.

C **Dusky Antechinus**

9 Projection and knob at about the same level.
10 Prominent flange.
11 Shaft longer and more robust than in Brown Antechinus (D).

Similar species: Other *Antechinus* species

D **Brown Antechinus**

12 Projection and knob at about the same level.
13 Flange not as prominent as in pygmy-possum (E).
14 Narrow shaft, long relative to size of ends.

Similar species: Other *Antechinus* species. Also dunnarts, planigales and other small insectivorous marsupials, and pygmy-possums (E).

E **Eastern Pygmy-possum**

15 Projection and knob at about the same level.
16 Prominent flange with sharp point.
17 Narrow shaft, long relative to size of ends.

Similar species: Western Pygmy-possum, Little Pygmy-possum, Long-tailed Pygmy-possum, Mountain Pygmy-possum, Feathertail Glider. Also some *Antechinus* species (D).

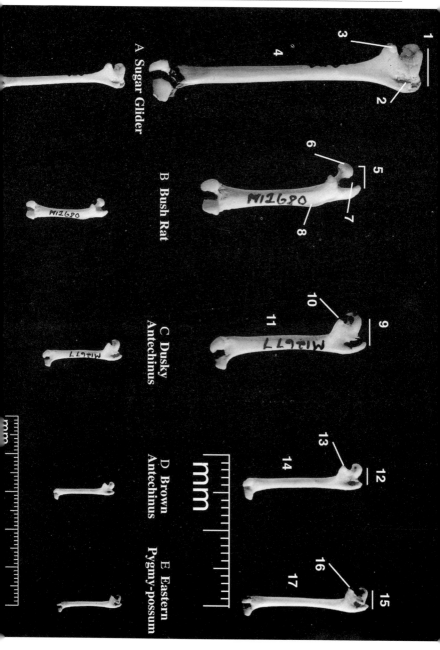

A Sugar Glider

B Bush Rat

C Dusky Antechinus

D Brown Antechinus

E Eastern Pygmy-possum

NON-MAMMALIAN BONES

The skulls and bones of other vertebrate animals can be mistaken for mammal bones. A few of the more commonly found non-mammalian bones are illustrated here.

Figure 245
Skull of parrot (see also page 270)

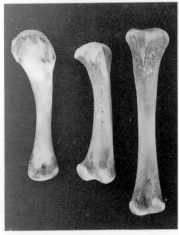

Figure 246
Humeri of three birds

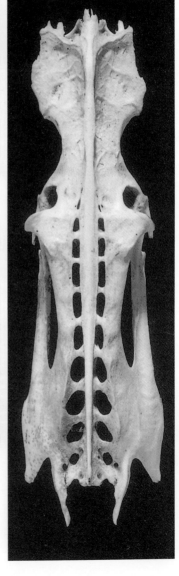

Figure 247
Keel of water-bird

330

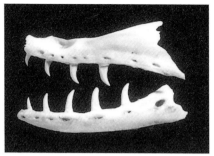

Figure 248   Teeth of goanna

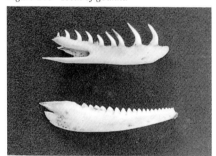

*Figure 249   Teeth of python (above)
and water dragon (below)*

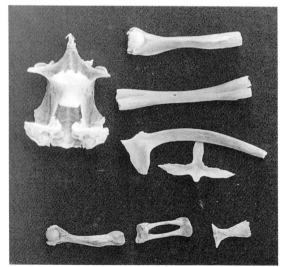

Figure 250   Amphibian bones, Cane toad (above) and frog (below)

BIBLIOGRAPHY

The following books are recommended for further reading.

Archer, M. (ed.) 1982. *Carnivorous Marsupials*. Royal Zoological Society of New South Wales.

Augee, M.L. & Gooden, B. 1993. *Echidnas of Australia and New Guinea*. University of New South Wales Press, Sydney.

Bang, P. & Dahlstrom, P. 1974. *Collins Guide to Animal Tracks and Signs*. Collins, London.

Brazenor, C.W. 1950. *The Mammals of Victoria*. Brown, Prior, Anderson Pty Ltd, Melbourne.

Brunner H. & Coman B.J. 1974. *The Identification of Mammalian Hair*. Inkata Press, Melbourne.

Burrell, H. 1974. *The Platypus*. Rigby, Adelaide.

Corbett, L.K. 1995. *The Dingo in Australia and Asia*. University of New South Wales Press, Sydney.

Dawson, T.J. 1995. *Kangaroos: The Biology of the Largest Marsupial*. University of New South Wales Press, Sydney.

Ellis M. & Etheridge, A. 1993. *Atlas of New South Wales Wildlife: Monotremes & Marsupials*. N.S.W. National Parks & Wildlife Service, Hurstville.

Forrest L.R. 1988. *Field Guide to Tracking Animals in Snow*. Stackpole Books, Harrisburg, PA.

Frith, H.J. & Calaby, J.H. 1969. *Kangaroos*. Cheshire, Melbourne.

Grant, T.R. 1984. *The Platypus*. University of New South Wales Press, Sydney.

Green, R.H. 1973. *The Mammals of Tasmania*. The author, Launceston.

Grigg, G., Jarman, J. & Hume, I. (eds.) 1989. *Kangaroos, Wallabies, Rat-kangaroos*. Surrey Beatty & Sons, Chipping Norton, New South Wales.

Hall, L.S. & Richards, G.C. 1979. *Bats of Eastern Australia*. Queensland Museum, Fortitude Valley, Queensland.

Ingram, G.J. & Raven, R.R. (eds) 1991. *An Atlas of Queensland's Frogs, Reptiles, Birds & Mammals*. Queensland Museum, Brisbane.

Lawrence, M.J. & Brown, R.W. 1973. *Mammals of Britain: Their Tracks, Trails and Signs*. Blandford Press Ltd., Poole, Dorset.

Lee, A.K. & Martin, R.W. 1988. *The Koala: A Natural History*. University of New South Wales Press, Sydney.

Lee, A.K., Handasyde, K.A. & Sanson, G.D. (eds) 1991. *Biology of the Koala*. Surrey Beatty & Sons, Chipping Norton, New South Wales

Lindenmayer, D.B., Tanton, M.T. & Norton, T.W. 1992. *Identification of the Forest Habitats of the Possums and Gliders of Central Victoria*. The authors, Australian National University, Canberra.

Lockley, R.M. 1964. *The Private Life of the Rabbit*. Andre Deutsch, London.

Mansergh, I.M. & Broome, L. 1994. *The Mountain Pygmy-possum of the Australian Alps*. University of New South Wales Press, Sydney.

Marlow, B.J. 1962. *Marsupials of Australia*. Jacaranda Press, Brisbane.

Menkhorst, P.W. (ed.) 1995. *Mammals of Victoria*. Oxford University Press, Melbourne.

Menkhorst, P.W. & Knight, F. 2001. *A Field Guide to the Mammals of Australia*. Oxford University Press, Melbourne.

Merrilees, D. & Porter, J.K. 1979. *Guide to the Identification of Teeth and Some Bones of Native Land Mammals Occurring in the Extreme South West of Western Australia*. Western Australian Museum, Perth.

Morrison, R.G.B. 1981. *A Field Guide to the Tracks and Traces of Australian Animals*. Rigby, Adelaide.

Murie, O.J. 1975. *A Field guide to Animal Tracks*. Houghton Mifflin Company, Boston.

Ride, W.D.L. 1970. *A Guide to the Native Mammals of Australia*. Oxford University Press, Melbourne.

Seebeck, J.H., Brown, P.R., Wallis, R.L. & Kemper, C.M. (eds) 1990. *Bandicoots and Bilbies*. Surrey Beatty & Sons, Chipping Norton, New South Wales.

Smith, A.P. & Hume, I.D. (eds) 1984. *Possums and Gliders*. Surrey Beatty & Sons, Chipping Norton, New South Wales.

Strahan, R. 2002 (Revised edition) *The Mammals of Australia*. Reed New Holland, Sydney.

Stonehouse, B. & Gilmore, D. (eds) 1977. *The Biology of Marsupials*. Macmillan, London.

Triggs, B.E. 1984. *Mammal Tracks and Signs: A Field Guide for South-eastern Australia*. Oxford University Press, Melbourne.

Triggs, B.E. 1988. *The Wombat: Common Wombats in Australia*. University of New South Wales Press, Sydney.

Tyndale-Biscoe, C.H. 1973. *Life of Marsupials*. Edward Arnold, London.

Wakefield, N.A. 1967. *A Naturalist's Diary*. Longman, Melbourne.

Walton, D.W. & Richardson, B.J. (eds) 1989. *Fauna of Australia. Volume 1B (Mammalia)*. Australian Government Publishing Service, Canberra.

Watts, C.H.S. & Aslin, H.J. 1981. *The Rodents of Australia*. Angus & Robertson, Melbourne.

Watts, D. 1993. *Tasmanian Mammals: A Field Guide*. Peregrine Press, Kettering, Tasmania

INDEX

Numbers in italics indicate illustrations

Acrobates pygmaeus, see Glider, Feathertail
Aepyprymnus rufescens, see Bettong, Rufous
Amphibians: bones *331*; tracks 83, *84*; scats
    185, *185*
Antechinomys laniger, see Kultarr
Antechinus
    all species: feeding signs 223; feet 49; gaits
      54; scats 167; shelters 223; tracks 53,
      54
    Agile: distribution *113*
    Atherton: distribution 153
    Brown: distribution *113*; femur 328, *329*;
      humerus 318, *319*; lower jaw 308, *309*;
      scats *115*; skull 284, *285*
    Cinnamon: distribution 153
    Dusky: distribution *108*; femur 328, *329*;
      humerus 318, *319*; lower jaw 308, *309*;
      scats *110*; skull 284, *285*; tracks 54
    Fawn: distribution 153
    Rusty: distribution *113*
    Subtropical: distribution *113*
    Swamp: distribution 153
    Yellow-footed: distribution *113*; scats *115*
Antechinus
    adustus, see Antechinus, Rusty
    agilis, see Antechinus, Agile
    bellus, see Antechinus, Fawn
    flavipes, see Antechinus, Yellow-footed
    godmani, see Antechinus, Atherton
    leo, see Antechinus, Cinnamon
    minimus, see Antechinus, Swamp
    stuartii, see Antechinus, Brown
    subtropicus, see Antechinus, Subtropical
    swainsonii, see Antechinus, Dusky
Arctocephalus
    forsteri, see Fur-seal, New Zealand
    pusillus, see Fur-seal, Australian
Axis axis, see Chital
Axis porcinus, see Deer, Hog

Bandicoot
    all species: feeding signs 205, *216*, 221;
      feet 46; gaits 48; scats 166, 221;
      shelters 221; tracks 47, *47*
    Eastern Barred: distribution *100*; scats *102*
    Golden: distribution 100
    Long-nosed: distribution *100*; feeding signs
      216; femur 326, *327*; humerus 316,
      317, lower jaw 302, *303*; scats *102*,
      216, shelters *216*, 221; skull 276, *277*;
      tracks *24*

Northern Brown: distribution *100*; lower
    jaw 302, *303*; scats *102*; skull 276, *277*
    Rufous Spiny: distribution 100; shelters
      221
    Southern Brown: distribution *100*; scats
      102; tracks *47, 48*
    Western Barred: distribution 100
Bat
    all species: feeding signs 241; feet 79; gaits
      79; scats 177; shelters 241; tracks 79, *79*
    Common Bent-wing: distribution *124*;
      scats *126*
    Ghost: tracks 79, shelters 241; feeding
      signs 241
    Gould's Wattled: lower jaw 308, *309*; skull
      284, *285*
    see also Flying-fox
Bettong
    all species: feeding signs *199*, 205; feet 31;
      gaits 31; scats 160; shelters 205; tracks
      31, *31*
    Brush-tailed (Woylie): distribution *148*;
      feeding signs *199*; scats *150*
    Burrowing: distribution 148
    Rufous: distribution *137*; feeding signs
      205; femur 324, *325*; humerus 316,
      317; lower jaw 300, *301*; scats *139*,
      160; skull 278, *279*; tracks *31*
    Northern: distribution 148; feeding signs
      199; scats *150*
    Tasmanian: distribution 148; scats *150*
Bettongia
    gaimardi, see Bettong, Tasmanian
    lesueur, see Bettong, Burrowing
    penicillata, see Bettong, Brush-tailed
    tropica, see Bettong, Northern
Bilby: distribution *104*; feeding signs *216*, 221;
    scats *106*, 166; shelters *216*, 221; tracks *24*,
    48
Birds: bones *330*; pellets *155*, *180*, *181*, *182*,
    243, *244*; scats 178, *179*; shelters 242;
    skull *270*, *330*; tracks *58*, 80–2
Bos
    javanicus, see Cattle, Banteng
    taurus, see Cattle, Domestic
Bubalus bubalis, see Buffalo, Water
Buffalo, Water: distribution 149; scats 174;
    tracks 75, *75*; wallow *223*, 239
Burramys parvus, see Pygmy-possum,
    Mountain

Camel, One-humped: distribution *125*; feet 73;
    gaits 73; scats *127*, 176; tracks *57*, 75, *75*

Camelus dromedarius, *see* Camel,
  One-humped
Canis
  familiaris, *see* Dog
  lupus dingo, *see* Dingo
  vulpes, *see* Fox, Red
Capra hircus, *see* Goat
Cat: distribution 96; feet 69, *69*; feeding signs
  238; scats *98*, 172, 238; shelters 237; skull
  264, *265*; tracks 57, 72, *72*
Cattle
  Banteng: distribution 149
  Domestic: distribution 149; feet 73; feeding
    signs *232*, 239; gaits 73; lower jaw 252,
    253; scats 176, *176*; shelters 239; skull
    252, *253*; tracks 75, *75*
Cercartetus
  caudatus, *see* Pygmy-possum, Long-tailed
  concinnus, *see* Pygmy-possum, Western
  lepidus, *see* Pygmy-possum, Little
  nanus, *see* Pygmy-possum, Eastern
Cervus
  elaphus, *see* Deer, Red
  timoriensis, *see* Deer, Rusa
  unicolor, *see* Deer, Sambar
Chalinolobus gouldii, *see* Bat, Chocolate
  Wattled
Chital: distribution 149
  see also Deer, all species
Conilurus penicillatus, *see* Rabbit-rat, Brush-
  tailed
Crab, *see* Invertebrates
Cuscus
  all species: feet 36; gaits 36; tracks 36
  Common Spotted: distribution *101*; scats
    103, 161; shelters 206
  Southern Common: distribution *101*; scats
    161; shelters 206

Dactylopsila trivirgata, *see* Possum, Striped
Dama dama, *see* Deer, Fallow
Dasycercus cristicauda, *see* Mulgara
Dasykaluta rosamondae, *see* Kaluta, Little Red
Dasyuroides byrnei, *see* Kowari
Dasyurus
  geoffroii, *see* Quoll, Western
  hallucatus, *see* Quoll, Northern
  maculatus, *see* Quoll, Spot-tailed
  viverrinus, *see* Quoll, Eastern
Deer
  all species: antlers 239, *240*; feet 3, *3*, 73,
    73; gaits 73; rubbiing trees *233*, 239;
    scats 174; shelters 239; tracks 76, *76*;
    wallows 239
  Fallow: distribution *149*; scats *151*
  Hog: distribution *149*; scats *151*
  Red: distribution *149*; scats *151*
  Rusa: distribution *152*; scats *154*
  Sambar: distribution *149*; scats *151*
Dendrologus
  bennettianus, *see* Tree-kangaroo, Bennett's
  lumholtzi, *see* Tree-kangaroo, Lumholtz's
Devil, Tasmanian: distribution 96; feeding
  signs 222; feet 49; gaits 52; lower jaw 290,

291; scats *98*, 166, 167; shelters 222; skull
  268, *269*; tracks 50, *50*, 52, *52*
Dibbler
  Southern: distribution 153
  Northern: distribution 153
Dingo: distribution *92*; feeding signs 237–8;
  feet *3*; femur 320, *321*; humerus 310, *311*;
  lower jaw 290; scats *94*, 172; shelters 237;
  skull 264, *265*; tracks *57*, 70
Dog: distribution *92*; feeding signs 237–8; feet
  3, 69, *69*; femur 320; gaits 70; humerus
  310; lower jaw 290; scats *94*, 172, *173*;
  shelters 237; skull 264; tracks 70, *70*; urine
  marking *231*, 238
Dockey: distribution 149; feet 73; scats 174,
  174; tracks 74
Dunnart
  all species: feet 55; gaits 55; scats 168;
    tracks 55, *55*
  Common: distribution *116*; scats *118*;
    shelters 223
  Fat-tailed: distribution *116*; scats *118*;
    shelters 223; tracks *55*
  Red-cheeked: distribution *116*; scats *118*
  White-footed: distribution *116*; scats *118*,
    168; shelters *218*, 223

Echidna, Short-beaked: distribution 93; feeding
  sign 193, *197*; feet 10, *10*; gaits 10; lower
  jaw 270, *271*; mating trench 194, *197*; scats
  95, 156; shelters 193; skull 270, *271*;
  tracks 11, *11*, *21*
Echymipera rufescens, *see* Bandicoot, Rufous
  Spiny
Equus
  asinus, *see* Donkey
  caballus, *see* Horse

Feeding signs: key 192
Feet: structure 3, *3*
Felis catus, *see* Cat
Femur: identification *247*, 250; key 251
Field-rat, Pale: distribution *112*; scats *114*;
  shelters 225
  see also Rat, all species
Flying-fox
  all species: feeding signs *234*, 241; scats 177
  Grey-headed: lower jaw 304, *305*; skull
    280, *281*
Fox, Red: distribution *92*; feeding signs 237–8,
  237; feet 69, *69*; gaits 70; lower jaw 290,
  291; scats *94*, 172, *173*; shelters *231*, 237;
  skull 264, *265*; tracks 71, *71*
Fur-seal
  all species: feet 77, *77*; gait 78; scats 177
  Australian: distribution *93*; lower jaw
    290, *291*; scats *95*; skull 264, *265*;
    tracks 78
  New Zealand: distribution 93

Gaits, 4–6, *5*, *6*
Glider
  all species: feet 34; gaits 38; scats 163–4;
    shelters 208; tracks 38–9, *38*, *39*

Feathertail: distribution *124*; gaits 41; scats *126*, 164; shelters 210; tracks 41, *41*

Greater: distribution *105*; femur 324, *325*; humerus 314, *315*; lower jaw 304, *305*; scats *107*, 163; shelters 208; skull 280, *281*; tracks 39

Mahogany: distribution 153; scats 163; tracks 39

Squirrel: distribution *113*; feeding signs 208; scats *115*, 163; shelters 208; tracks 39

Sugar: distribution *113*; feeding signs 208; femur 328, *329*; humerus 318, *319*; lower jaw 306, *307*; scats *115*, 163; shelters *201*, 208; skull 282, *283*; tracks *38*, 39, *39*

Yellow-bellied: distribution *105*; feeding signs *203*, *204*, 208; scats *107*, 163; shelters 208; tracks 39

Goannas, *see* Reptiles

Goat: distribution *152*; gaits 73; lower jaw 286, *287*; scats *154*, 174, *175*; shelters 239; skull 254, *255*; tracks 76, *76*

Gymnobelideus leadbeaterii, *see* Possum, Leadbeater's

Hair 89

Hare, Brown: distribution *152*; feeding signs 228; feet 67; gaits 68; scats *154*, 171; shelters 228; tracks *67*, 68

Hare-wallaby
all species; feet 25; scats 159; shelters 196; tracks 30

Banded: distribution 145

Rufous (Mala): distribution *145*; scats *147*

Spectacled: distribution *145*; lower jaw 298, 299; scats *147*; skull 274, *275*; tracks 25, 30, *30*

Hemibelideus lemuroides, *see* Possum, Lemuroid Ringtail

Honey Possum: distribution 124; feet 41; gaits 41; scats 164; shelters 210

Hopping-mouse
all species; feet 62; gaits 66; scats 170; shelters 226; tracks 66, *66*

Dusky: distribution *121*; scats *123*

Fawn: distribution 121

Mitchell's: distribution *121*; scats *123*

Northern: distribution 121

Spinifex: distribution *121*; scats *123*; tracks *66*

Horse: distribution 149; feet 3, *3*; gaits 73; lower jaw 252, *253*; scats 174; skull 252, *253*; tracks 74, *74*

Horseshoe-bat, Eastern: distribution *124*; scats *126*

Humerus: identification *247*, 248; key 251

Hydromys chrysogaster, *see* Water-rat

Hydrurga leptonyx, *see* Seal, Leopard

Hypsiprymnodon moschatus, *see* Rat-kangaroo, Musky

Invertebrates: feeding signs 242; scats 186, *187*; shelters *236*, 242; tracks 84, *84*

Isoodon
auratus, *see* Bandicoot, Golden
macrourus, *see* Bandicoot, Northern Brown
obesulus, *see* Bandicoot, Southern Brown

Kaluta, Little Red: distribution 153

Kangaroo
all species: feet 3, *3*, 14; feeding signs 195; gaits 15, *15*; scats 157; shelters 195; scratch marks 195; tracks *2*, 15–20; water-digs 195, *198*

Eastern Grey: distribution *129*; feet 17; femur 320; humerus 310, *311*; lower jaw 288, *289*; scats *131*, 157, *157*, *158*; skull 256, *257*; tracks *16*, *17*, 18, *18*, *21*, *22*

Red: distribution *129*; feet 18; scats *131*; tracks *18*, 21

Western Grey: distribution *129*; feet 18; scats *131*; tracks 18

Koala: distribution *101*; feet 42; femur 322, *323*; gaits 43; humerus 312, *313*; lower jaw 292, *293*; scats *103*, 165; scratch marks 212, *213*; shelters 211; skull 268, *269*; tracks 43, *43*

Kowari: distribution *108*; scats *110*, 167; shelters *218*, 222; tracks *24*, 52, *52*

Kultarr: distribution 116; gaits 55; shelter 223

Lagostrophus fasciatus, *see* Hare-wallaby, Banded

Lagorchestes
conspicillatus, *see* Hare-wallaby, Spectacled
hirsutus, *see* Hare-wallaby, Rufous

Lasiorhinus
krefftii, *see* Wombat, Northern Hairy-nosed
latifrons, *see* Wombat, Southern Hairy-nosed

Leggadina
forresti, *see* Mouse, Forrest's
lakedownensis, *see* Mouse, Lakeland Downs

Leporillus
apicalis, *see* Stick-nest Rat, Lesser
conditor, *see* Stick-nest Rat, Greater

Lepus capensis, *see* Hare, Brown

Lizards, *see* Reptiles

Lobodon carcinophagus, *see* Seal, Crab-eater

Lower jaw: identification *247*, 248; key 251

Macropus
agilis, *see* Wallaby, Agile
antilopinus, *see* Wallaroo, Antilopine
bernardus, *see* Wallaroo, Black
dorsalis, *see* Wallaby, Black-striped
eugenii, *see* Wallaby, Tammer
fuliginosus, *see* Kangaroo, Western Grey
giganteus, *see* Kangaroo, Eastern Grey
irma, *see* Wallaby, Western Brush
parma, *see* Wallaby, Parma
parryi, *see* Wallaby, Whiptail
robustus, *see* Wallaroo, Common
rufogriseus, *see* Wallaby, Red-necked
rufus, *see* Kangaroo, Red

Macrotis lagotis, *see* Bilby
Mammals, hoofed: feet 73; shelters 239
    see also Cattle; Deer; Goat; Horse; Pig;
    Sheep
Marsupials mice, *see* Antechinus, all species
Mastocomys fuscus, *see* Rat, Broad-toothed
Melomys
    Cape York; distribution 112
    Fawn-footed: distribution *112*; scats *114*;
        shelters 226
    Grassland: distribution *112*; scats *114*;
        shelters 226
    see also Rat, all species
Melomys
    burtoni, *see* Melomys, Grassland
    capensis, *see* Melomys, Cape York
    cervinipes, *see* Melomys, Fawn-footed
Mesembriomys
    gouldii, *see* Tree-rat, Black-footed
    macrurus, *see* Tree-rat, Golden-backed
Miniopterus schreibersii, *see* Bat, Common
    Bent-wing
Mirounga leonina, *see* Seal, Southern Elephant
Mole, Marsupial: distribution 104; feet 61; gait
    61; shelters *219*, 222; tracks 61, *219*
Monjon: distribution 141
    see also Rock Wallaby, all species
Mouse
    all species: feet 62; gaits 63; tracks 63, *64*;
        shelters 225
    Ash-grey: distribution *120*; scats *122*
    Desert: scats 169
    Forrest's: distribution 153
    Heath: distribution *117*; scats *119*
    House: distribution *121*; odour 169, 227;
        scats *123*, 169; shelters 225; skull 282,
        283; tracks *64*
    Lakeland Downs; distribution 153
    Long-tailed: distribution *120*; scats *122*;
        shelters 225–6
    New Holland: distribution *124*; scats *126*;
        shelters 225
    Plains: distribution *120*; scats *122*
    Sandy Inland: distribution *121*; scats *123*,
        169; shelters *229*
    Silky: distribution *120*; scats *122*; shelters
        225
    Smoky: distribution *117*; scats *119*
    Western Pebble-mound: shelters 226, *229*
    see also Hopping-mouse
Mulgara: distribution *108*; scats *110*, 167;
    shelters *218*, 222; tracks 52
Mus musculus, *see* Mouse, House
Myrmecobius fasciatus, *see* Numbat

Nabarlek: distribution *145*; scats *147*
    see also Rock-wallaby, all species
Neophoca cinerea, *see* Sea-lion, Australian
Ningaui
    Mallee: distribution *116*; scats *118*
    Pilbara: distribution 120
    Wongai: distribution *120*; scats *122*
Ningaui
    ridei, *see* Ningaui, Wongai

*   timealeyi*, *see* Ningaui, Pilbara
    yvonneae, *see* Ningaui, Southern
Notomys
    alexis, *see* Hopping-mouse, Spinifex
    aquilo, *see* Hopping-mouse, Northern
    cervinus, *see* Hopping-mouse, Fawn
    fuscus, *see* Hopping-mouse, Dusky
    mitchelli, *see* Hopping-mouse, Mitchell's
Notoryctes typhlops, *see* Mole, Marsupial
Numbat: distribution *104*; feeding signs *219*,
    224; feet 61; gaits 61; scats *106*, 168;
    shelters 224; tracks *61*

Onychogalea
    fraenata, *see* Wallaby, Bridled Nailtail
    unguifera, *see* Wallaby, Northern Nailtail
Ornithorhynchus anatinus, *see* Platypus
Other traces: key 192
Ovis aries, *see* Sheep
Owl, *see* Birds

Pademelon
    all species: feet 25; feeding signs 196; gaits
        26; paths 196; scats 159; shelters 196;
        tracks 25–6, *26*
    Red-legged: distribution *140*; scats *142*;
        tracks *26*
    Red-necked
        distribution *140*; lower jaw 296,
        297; scats *142*; skull 272, *273*, 274, *275*
    Tasmanian: distribution *140*; scats *142*;
        tracks 25
Parantechinus
    apicalis, *see* Dibbler, Southern
    bilarni, *see* Dibbler, Northern
Perameles
    bougainville, *see* Bandicoot, Western Barred
    gunnii, *see* Bandicoot, Eastern Barred
    nasuta, *see* Bandicoot, Long-nosed
Petauroides volans, *see* Glider, Greater
Petaurus
    australis, *see* Glider, Yellow-bellied
    breviceps, *see* Glider, Sugar
    gracilis, *see* Glider, Mahogany
    norfolcensis, *see* Glider, Squirrel
Petrogale
    assimilis, *see* Rock-wallaby, Allied
    brachyotis, *see* Rock-wallaby, Short-eared
    burbidgei, *see* Monjon
    coenensis, *see* Rock-wallaby, Cape York
    concinna, *see* Nabarlek
    godmani, *see* Rock-wallaby, Godman's
    herberti, *see* Rock-wallaby, Herbert's
    inornata, *see* Rock-wallaby, Unadorned
    lateralis, *see* Rock-wallaby, Black-flanked
    mareeba, *see* Rock-wallaby, Mareeba
    penicillata, *see* Rock-wallaby, Brush-tailed
    persephone, *see* Rock-wallaby, Proserpine
    purpureicollis, *see* Rock-wallaby, Purple-
        necked
    rothschildi, *see* Rock-wallaby, Rothschild's
    sharmani, *see* Rock-wallaby, Sharman's
    xanthopus, *see* Rock-wallaby, Yellow-footed
Petropseudes
    dahli, *see* Possum, Rock Ringtail

Phalanger
intercastellanus, see Cuscus, Southern
Common
Phascogale
all species: feet 49; feeding signs 223; gaits
53; scats 167; tracks 50, 53, *53*
Brush-tailed: distribution *108*; lower jaw
306, *307*; scats *110*, 167; shelters 223;
skull 282, *283*; tracks *53*
Red-tailed: distribution *116*; scats *118*;
shelters 223
Phascogale
calura, see Phascogale, Red-tailed
penicillata, see Phascogale, Brush-tailed
Phascolarctos cinereus, see Koala
Pig: distribution *125*; feeding signs *232*, 239;
feet 73; gaits 73; lower jaw 286, *287*; scats
127, 176; shelters 239; skull 254, *255*;
tracks 76, *76*
Planigale
all species: feeding signs 223; feet 49;
scats 168; tracks 55
Common: distribution *120*; scats *122*
Giles': distribution 120
Long-tailed: distribution 120
Narrow-nosed: distribution 120
Planigale
gilesi, see Planigale, Giles'
ingrami, see Planigale, Long-tailed
maculata, see Planigale, Common
tenuirostris, see Planigale, Narrow-nosed
Platypus: distribution 105; feeding signs 194;
feet 12, *12*; gaits 12; lower jaw 270, *271*;
scats *107*, 156; shelters 194, *197*; skull
270, *271*; tracks 12, *13*
Pogonomys mollipilosus, see Rat, Prehensile-
tailed
Possum
Brushtail: all species: feet 34; feeding signs
200, 206; gaits 35; scats 161; scratch
marks 200, 206; shelters 206; tracks 1,
35, *35*
Common Brushtail: distribution *104*;
feet *34*; femur 326, *327*; humerus
314, *315*; lower jaw 300, *301*; scats
106, 161, *162*; shelters *200*, 206;
skull 278, *279*; tracks 35, *36*
Mountain Brushtail: distribution 101;
scats *103*, 161; shelters 206; tracks
36; Northern: distribution 104; scats
106, 161; shelter 206
Leadbeater's: distribution *113*; feeding
signs *202*, 208; feet 39; scats *115*, 163;
shelters *202*, 208; tracks 39
Ringtail: all species: feet 34, 37; feeding
signs 207; scats 163; scratch marks
207; tracks 37
Common Ringtail: distribution 105;
feeding signs *201*, 207; femur 326,
327; humerus 316, *317*; lower jaw
304, *305*; scats *107*, 163; shelters *201*,
207; skull 280, *281*; tracks 37, *37*
Daintree River Ringtail: distribution 105
Green Ringtail: distribution 105;
shelters 207

Herbert River Ringtail: distribution
105; shelters 207; scats *107*
Lemuroid Ringtail: distribution 105;
scats *107*
Rock Ringtail: distribution *104*; scats
106, 163
Western Ringtail: distribution 105
Scaly-tailed: distribution 101; feet 36; gaits
36; scats 161; shelters 206; tracks 36
Striped: distribution 153; feeding signs
209; feet 39; scats 164; shelters 208;
tracks 39
see also Honey Possum
Potoroo
all species: feeding signs *199*, 205; feet 31;
gaits 31; scats 160; shelters 205; tracks
32, *32*
Gilbert's: distribution 148
Long-footed: distribution *148*; scats *150*
Long-nosed: distribution *148*; feeding signs
199; feet 31; femur 326, *327*; humerus
316, *317*; lower jaw 302, *303*; scats
150; skull 278, *279*; tracks 22, *32*
see also Rat-kangaroo
Potorous
gilberti, see Potoroo, Gilbert's
longipes, see Potoroo, Long-footed
tridactylus, see Potoroo, Long-nosed
Pseudantechinus
Carpentarian: distribution 153
Fat-tailed: distribution 153
Little Red Kaluta: distribution 153
Ningbing: distribution 153
Tan: distribution 153
Woolley's: distribution 153
Pseudantechinus
macdonellensis, see Pseudantechinus,
Fat-tailed
mimulus, see Pseudantechinus, Carpentarian
ningbing, see Pseudantechinus, Ningbing
roryi, see Pseudantchinus, Tan
woolleyae, see Pseudantechinus, Woolley's
Pseudocheirus
occidentalis, see Possum, Western Ringtail
peregrinus, see Possum, Common Ringtail
Pseudochirops
archeri, see Possum, Green Ringtail
Pseudochirulus
cinereus, see Possum, Daintree River
Ringtail
herbertensis, see Possum, Herbert River
Ringtail
Pseudomys
albocinereus, see Mouse, Ash-grey
apodemoides, see Mouse, Silky
australis, see Mouse, Western Pebble-
mound
desertor, see Mouse, Desert
fumeus, see Mouse, Smoky
hermannsburgensis, see Mouse, Sandy
Inland
higginsi, see Mouse, Long-tailed
novaehollandiae, see Mouse, New Holland
shortridgei, see Mouse, Heath
Pterropus sp., *see* Flying-fox

Pygmy-possum
    all species: feet 40; gaits 40; scats 164;
        tracks 40
    Eastern: distribution *117*; femur 328, *329*;
        humerus 318, *319*; lower jaw 308, *309*;
        scats *119*; shelters *204*, 210; skull 284,
        285; tracks 40
    Little: distribution *124*; scats *126*; shelters 210
    Long-tailed: distribution 124; shelters 210
    Mountain: distribution *117*; feeding signs
        210, *210*; scats *119*, 164; shelters 210
    Western: distribution *124*; scats *126*;
        shelters 210

Quokka: distribution *136*; shelters 196; scats
    138, 159; tracks 30, *30*
    see also Wallaby, all species
Quoll
    all species: feet 49; gaits 51; scats 166;
        shelters 222; tracks 51
    Eastern: distribution 97; feet 49; scats *99*,
        166; shelters 222; tracks *24*, 51
    Northern: distribution 97; scats *99*, 167;
        shelters *217*, 222; tracks 51
    Spot-tailed: distribution 96; feet *49*; femur
        324, *325*; humerus 314, *315*; lower jaw
        300, *301*; scats *98*, 166, 167; shelters
        217, 222; skull 272, *273*; tracks *51*
    Western: distribution 97; scats *99*, 167;
        shelters 222; tracks 51

Rabbit, European: distribution *152*; feeding
    signs 228, *230*; feet 67, *67*; femur 324,
    325; gaits 68; humerus 314, *315*; lower jaw
    302, *303*; paths 228; scats *152*, 171, *171*,
    228, *228*; shelters 228, *230*; skull 276, *277*;
    tracks 68, *68*
Rabbit-rat, Brush-tailed: distribution *117*; scats
    119; shelters 226
    see also Rat, all species
Raptors, *see* Birds
Rat
    all species: feeding signs 226–7; feet 62;
        gaits 63; runways 86, *86*, 227; scats
        169; shelters 225; tracks 63, *63*
    Black: distribution *109*; scats *111*, 169;
        shelters *220*, 225; tracks 64, *64*
    Broad-toothed: distribution *109*, scats *111*,
        169; shelters *220*, 225
    Brown: distribution *109*; scats *111*; shelters
        225
    Bush: distribution *109*; feeding signs 210,
        210; feet *62*; femur 328, *329*; humerus
        318, *319*; lower jaw 306, *307*; scats
        111, 169; shelters 225; skull 282, *283*;
        tracks *63*
    Canefield: distribution *109*; scats *111*;
        shelters 225
    Cape York: distribution 112
    Dusky: distribution *112*; scats *114*; shelters
        225
    Giant White-tailed: distribution *112*; scats
        114; shelters 226
    Long-haired: distribution 112; shelters *220*,
        225

Masked White-tailed: distribution *112*;
    scats *114*
Prehensile-tailed: distribution 121
Swamp: distribution *109*; runways *86*; scats
    111; shelters 225
see also Melomys; Stick-nest Rat; Rock-
    rat; Tree-rat; Water-rat; Water-rat, False
Rattus
    colletti, see Rat, Dusky
    fuscipes, see Rat, Bush
    leucopus, see Rat, Cape York
    lutreolus, see Rat, Swamp
    norvegicus, see Rat, Brown
    rattus, see Rat, Black
    sordidus, see Rat, Canefield
    tunneyi, see Field-rat, Pale
    villosissimus, see Rat, Long-haired
Rat-kangaroo, Musky: distribution *113*; gaits
    31, 33; shelters 205; scats *115*, 160; tracks
    33, *33*
    see also Potoroo
Reptiles: bones 331; gaits 83; scats 183, *183*,
    184; shelters *236*, 242; tracks *59*, *60*, 83,
    83
Rhinolophus megaphyllus, see Horseshoe-bat,
    Eastern
Rock-rat
    all species: shelters 226
    Arnhem Land: distribution 153
    Carpentarian: distribution 153
    Central: distribution 153
    Common: distribution *117*; scats *119*
    Large: distribution 153
Rock-wallaby
    all species: feeding signs 196; feet 25, 27;
        paths 196; scats 159; shelters 196;
        tracks 27
    Allied: distribution *144*; scats *146*
    Black-footed: distribution 141
    Brush-tailed: distribution *141*; lower jaw
        296, *297*; scats *143*; skull 274, *275*
    Cape York: distribution 144
    Godman's: distribution 144
    Herbert's: distribution 144
    Mareeba: distribution 144
    Purple-necked: distribution *144*; scats *146*
    Proserpine: distribution *144*; scats *146*
    Rothschild's: distribution 141
    Sharman's: distribution 144
    Short-eared: distribution 141
    Unadorned: distribution *141*; scats *143*;
        tracks 27
    Yellow-footed: distribution 141; feet 27;
        scats *143*
    see also Monjon; Nabarlek

Sarcophilus harrisii, see Devil, Tasmanian
Scats
    analysis 88–9, *88*, *89*
    identification 87–8
    key 90–1
Seal
    all species: feet 77; gaits 78; scats 177
    Crab-eater: distribution 93
    Leopard: distribution 93

Southern Elephant: distribution 93; feet 78; gaits 78

Sea-lion, Australian: distribution 93; feet 78; gaits 78; tracks *57*, 78, *78*

Setonix brachyurus, see Quokka

Sheep: distribution *152*; gaits 73; lower jaw 286, *287*; scats *154*, 174, *175*; skull 254, *255*; tracks 76, *76*

Shelters: key 190–1

Skulls

identification 246, *247*, 248

key 251

Sminthopsis

crassicaudata, see Dunnart, Fat-tailed

leucopus, see Dunnart, White-footed

murina, see Dunnart, Common

virginiae, see Dunnart, Red-cheeked

Snakes, *see* Reptiles

Stick-nest Rat

all species: shelters 226, *229*

Greater: distribution 153

Lesser: status 226

see also Rat, all species

Sus scrofa, see Pig

Tachyglossus aculeatus, see Echidna, Short-beaked

Tarsipes rostratus, see Honey Possum

Thylacine: feet 56; gaits 56; status 49; tracks 56, *56*

Thylacine cynocephalus, see Thylacine

Thylogale

billardierii, see Pademelon, Tasmanian

stigmatica, see Padelmon, Red-legged

thetis, see Padmelon, Red-necked

Tracks 1–2

key 7–9

Tree-kangaroo

all species: feet 29; gaits 29; scats 159; scratch marks 196; shelters 196; tracks 29, 196

Bennett's: distribution 137

Lumholtz's: distribution *137*; lower jaw 298, *299*; scats *139*; skull 274, *275*; tracks *29*

Tree-rat

Black-footed: distribution 153; shelters 226

Golden-backed: distribution 153; shelters 226

see also Rat, all species

Trichosurus

caninus, see Possum, Mountain Brushtail

vulpecular, see Possum, Common Brushtail

Uromys

caudimaculatus, see Rat, Giant White-tailed

hadrourus, see Rat, Masked White-tailed

Vombatus ursinus, see Wombat, Common

Vulpes vulpes, see Fox, Red

Wallabia bicolour, see Wallaby, Swamp

Wallaby

all species: feet 14; feeding signs 195, *198*; gaits 15, *15*; paths 195; scats 157, 159; shelters 195; tracks 15–20

Agile: distribution *133*; lower jaw 298, *299*; scats *135*; tracks *19*

Black-striped: distribution *136*; scats *138*; tracks *19*

Nailtail: all species: shelters 196; Bridled: distribution *137*; scats *139*; Northern: distribution *137*; feet 28; lower jaw 296, *297*; scats *139*; skull 274, *275*; tracks 25, 28, *28*

Parma: distribution *136*; scats *138*; tracks *19*

Red-necked: distribution *132*; lower jaw 294, *295*; scats *134*, *158*; skull 258, *259*; tracks *16*

Swamp: distribution *132*; feet *14*; femur 320, *321*; humerus 310, *311*; lower jaw 294, *295*; scats *134*, 157, *158*; skull 260, *261*; tracks *17*

Tammar: distribution *136*; scats 138; tracks *19*

Western Brush: distribution 136; tracks *19*

Whiptail: distribution *132*; scats *134*, *158*; tracks *19*

see also Hare-wallaby; Rock-wallaby

Wallaroo

all species: feet 14; feeding signs 195; gaits 15; scats 157; shelters 195; tracks 15–18; water-digs 195

Antilopine: distribution *133*; scats *135*; tracks *18*

Black: distribution 133; tracks *18*

Common: distribution *133*; scats *135*, *158*; tracks *18*

Water-rat: distribution *108*; feeding signs *170*, 227, *230*; feet 62, 65; gaits 65; scats *110*, 170; shelters 226; tracks 65, *65*

Water-rat, False: distribution 108

see also Rat, all species

Wombat

all species: feeding signs 212; feet 42; gaits 44; rubbing posts 212, *215*; paths 212, *214*; scats 165; shelters 211; tracks 44–5; wallows 212

Common: distribution *128*; feeding signs 212; feet *42*; femur 322, *323*; humerus 312, *313*; lower jaw 292, *293*; scats *85*, *130*, 165; scratch marks 212, *215*; shelters 211, *213*, *214*; skull *257*, 266; teeth *266*; tracks *23*, 44, *45*

Northern Hairy-nosed: distribution *128*; scats *130*, 165; shelters 211, *214*; tracks *23*

Southern Hairy-nosed: distribution *128*; scats *130*, 165; shelters 211, *214*

Xeromys myoides, see Water-rat, False

Yabbie, *see* Invertebrates

Zyzomys

argurus, see Rock-rat, Common

maini, see Rock-rat, Arnhem Land

palatalis, see Rock-rat, Carpentarian

pedunculatus, see Rock-rat, Central

woodward, see Rock-rat, Large